Laboratory Manual

to accompany

Chemistry in Context
Applying Chemistry in Society

Fourth Edition

Edited by

Wilmer J. Stratton
Earlham College

and

Gail A. Steehler
Roanoke College

Norbert J. Pienta
University of Iowa

Catherine H. Middlecamp
University of Wisconsin–Madison

Boston   Burr Ridge, IL   Dubuque, IA   Madison, WI   New York   San Francisco   St. Louis
Bangkok   Bogotá   Caracas   Kuala Lumpur   Lisbon   London   Madrid   Mexico City
Milan   Montreal   New Delhi   Santiago   Seoul   Singapore   Sydney   Taipei   Toronto

The McGraw·Hill Companies

Laboratory Manual to accompany
CHEMISTRY IN CONTEXT: APPLYING CHEMISTRY TO SOCIETY, FOURTH EDITION
STRATTON, STEEHLER, PIENTA, MIDDLECAMP

Published by McGraw-Hill Higher Education, an imprint of The McGraw-Hill Companies, Inc.,
1221 Avenue of the Americas, New York, NY 10020. Copyright © The McGraw-Hill Companies,
Inc., 2003, 2000, 1997, 1994. All rights reserved.

Some of the laboratory experiments included in this text may be hazardous if materials are handled
improperly or if procedures are conducted incorrectly. Safety precautions are necessary when you
are working with chemicals, glass test tubes, hot water baths, sharp instruments, and the like, or for
any procedures that generally require caution. Your school may have set regulations regarding safety procedures that your instructor will explain to you. Should you have any problems with materials
or procedures, please ask your instructor for help.

  This book is printed on recycled paper containing 10% postconsumer waste.
RECYCLED

1 2 3 4 5 6 7 8 9 0 QPD/ QPD 0 9 8 7 6 5 4 3 2

ISBN 0-07-242457-5

Publisher: *Kent A. Peterson*
Sponsoring editor: *Kent A. Peterson*
Developmental editor: *Shirley R. Oberbroeckling*
Marketing manager: *Thomas D. Timp*
Project manager: *Joyce Watters*
Lead production supervisor: *Sandy Ludovissy*
Designer: *Stuart Paterson*
Media producer: *Jeffry Schmitt*
Compositor: *Terri Schiesl*
Printer: Quebecor World Dubuque, IA

www.mhhe.com

Contents

Preface to Instructors

This laboratory manual is intended to accompany the fourth edition of *Chemistry in Context, Applying Chemistry to Society*. That text and this laboratory manual have been designed specifically for college students majoring in disciplines outside of the natural sciences. The text examines a selected set of science and technology issues with broader societal implications; this manual provides laboratory experiences that are relevant to those topics. The authors believe that laboratory work ought to be an integral part of chemistry courses for this student population. Hands-on experience with experimentation and data collection are crucial to an understanding of the scientific method and to the role that science plays in addressing societal issues.

In general, the experiments are relatively simple with easy-to-follow instructions. Some are adaptations of traditional experiments; others are quite novel. Many use small-scale equipment of a type that has become popular in recent years. (More discussion of this equipment and its use can be found in the *Instructor's Resource Guide*.) The experiments require very little mastery of traditional laboratory techniques; thus, maximum student time can be devoted to explorations and acquiring data.

Most of the experiments can be completed in approximately 1 hour. A few are considerably longer. For courses with a 2-hour or 3-hour lab period, it is possible to do two experiments in one period or to utilize the suggestions included with many experiments for optional extensions or alternate assignments. For courses that do not have a scheduled lab component, it is possible to do some of the experiments in a classroom setting with fairly simple modifications, especially those experiments using small-scale equipment.

Most experiments from the third edition have been retained. Any significant modifications of those experiments are highlighted in the *Instructor's Resource Guide*. Six new experiments have been added (Experiments 2, 4, 5, 21, 22, and 23), mostly to fill gaps related to topics in the text. To make room for these additions, five experiments from the third edition have been moved to the Chemistry-in-Context website where they can be downloaded for class use if desired. (Go to www.mhhe.com/cic and look under Instructor Resources.) Also new to this edition is a set of "Performance-Based Assessment" activities, given at the end of the laboratory manual. More detail about these and ways to use them can be found in the online resources.

An *Instructor's Resource Guide* is available online (www.mhhe.com/cic, under Instructor Resources) that provides valuable information for each of the experiments in this manual. Instructors and lab support personnel are strongly urged to make use of this resource. The guide includes lists of materials, pedagogical objectives, teaching hints, trouble shooting, possible extensions, sample student results, and answers to questions.

This collection of experiments is the result of a collaborative effort involving many people. Major contributions to earlier editions came from Robert Silberman, Truman Schwartz, and Conrad Stanitski. For the new edition, a number of users contributed generously with ideas and willingness to do class testing of new or revised experiments. We especially wish to thank Donald Phillips at Eastern Michigan University, Jack Steehler at Roanoke College, and Linda Pallack at Washington & Jefferson College.

Finally, a word about the status of these experiments. Most have been used by *Chemistry in Context* authors (past and present) in our own CiC classes. They are versions that we have found to fit our needs. But each course is different, and instructors have differing styles. We invite users to contact us with suggestions for modifications of these experiments or for exchange of ideas about possible new experiments.

To the Student

The experiments in this text have been carefully chosen and designed to reflect and amplify the contents of *Chemistry in Context*. Our object in writing this manual was not to train future chemists, but rather to show what chemists do and how they do it. We also would like for you to appreciate the fact that there is no great mystery to doing chemistry in the laboratory. A great deal of chemical information about the world around you can be obtained with simple chemical equipment and simple procedures.

Another goal is to give you an opportunity to discover how chemists solve problems. After all, chemistry is an experimental science, and most chemists spend part of their time in a laboratory. Like practicing chemists, you will use the laboratory to try out new ideas, investigate the properties of materials and compounds, synthesize compounds, analyze materials, and, in general, solve problems. The experiments in this text show you how it is done. Two of the experiments, *Hot Stuff: An Energy Conservation Problem* and *Solubilities: An Investigation*, ask you to solve problems as a chemist might when no instructions are available. In addition, various optional challenges or extensions are scattered throughout the manual.

You will find that for most of these experiments quite simple equipment is used and that you can easily learn the necessary techniques. In general, you will be working with a partner, and in some cases, the whole class will work collaboratively to collect data and answer a scientific question.

For nearly every experiment, you will find one or more data sheets at the end. Unless your instructor directs you otherwise, you should record all observations and data on these sheets. *It is important to make these records while you are doing the experiment, not at some later time when you may have forgotten the details.* Each experiment also includes a set of questions to be answered after completing the experiment. These questions are designed to help you consolidate what you have learned and to demonstrate your understanding.

Although you may never again work in a chemistry laboratory, it is our hope that after this laboratory course, you will understand why chemists find laboratory work so interesting and compelling.

Essay: The Wonder of Chemistry Experiments

The quote below, from F.H. Getman's article on the life of Ira Remsen,* has often been cited by chemists in lectures and texts—perhaps because it reminds them of some similar incident that happened to them during their adolescent years and influenced their decision to become chemists. We quote it because it conveys, in a charming way, the message that you can't really know some aspects of chemistry unless you experience them for yourself   While we certainly hope you will come away from the lab with your clothes and skin intact, we would also like you to come away with some of the wonder and interest in chemistry that was kindled in Ira Remsen by his first laboratory experience. Remsen went on to become one of the foremost American chemists of the nineteenth century.

> While reading a textbook of chemistry, I came across the statement, "nitric acid acts on copper." I was getting tired of reading such absurd stuff and I was determined to see what this meant. Copper was more or less familiar to me, for copper cents were then in use. I had seen a bottled marked nitric acid on a table in the doctor's office where I was then "doing time." I did not know its peculiarities, but the spirit of adventure was upon me. Having nitric acid and copper, I had only to learn what the words "act upon" meant. The statement "nitric acid acts upon copper" would be something more than mere words. All was still. In the interest of knowledge I was even willing to sacrifice one of few copper coins then in my possession. I put one of them on the table, opened the bottle marked nitric acid, poured some of the liquid on the copper and prepared to make an observation. But what was this wonderful thing which I beheld? The cent was already changed and it was no small change either. A green-blue liquid foamed and fumed over the cent and over the table. The air in the neighborhood of the performance became colored dark red. A great colored cloud arose. This was disagreeable and suffocating. How should I stop this? I tried to get rid of the objectionable mess by picking it up and throwing it out the window. I learned another fact. Nitric acid not only acts on copper, but it acts upon fingers. The pain led to another unpremeditated experiment. I drew my fingers across my trousers and another fact was discovered. Nitric acid acts upon trousers. Taking everything into consideration, that was the most impressive experiment and relatively probably the most costly experiment I have ever performed.... It was a revelation to me. It resulted in a desire on my part to learn more about this remarkable kind of action. Plainly, the only way to learn about it was to see its results, to experiment, to work in a laboratory.

* From *The Life of Ira Remsen*, published by the *Journal of Chemical Education*, of the Division of Chemical Education Inc., of the American Chemical Society.

Some Notes About Laboratory Safety

Good laboratory practice requires that you take some simple safety precautions whenever you work in a chemistry laboratory. The popular notion that a chemistry lab is a dangerous place, filled with unknown disasters waiting to occur, is simply untrue for most situations and is certainly incorrect for the activities in this laboratory text. Nevertheless, all chemistry laboratories have some potential hazards associated with chemical splashes, careless handling of flammable substances, and glassware breakage. With this in mind, what follows are some basic rules that you must adhere to when working in a chemistry laboratory.

1. **Wear approved eye protection at all times**. This is the most essential rule and will be rigidly enforced. Chemical splashes are most serious with regard to your eyes. Even dilute solutions of many chemicals can cause serious eye damage upon contact. Fortunately, your eyes are very easy to protect by wearing approved safety glasses or goggles. Even when you are not working directly with chemicals, someone near you may have an accident and something may splash or fly in your direction.

2. **Exercise special care whenever you are using flammable chemicals.** There must be no open flames anywhere in the vicinity of your work. Even a <u>hot</u> <u>object</u> can sometimes cause flammable vapors to ignite. This means that you must know which liquids are flammable.

3. **Never eat, drink, or smoke in the laboratory**. Some chemicals are very toxic.

4. **Never perform unauthorized experiments**. Some simple chemicals can form explosive or toxic products when mixed in unintended or inappropriate ways.

5. **Never work in a laboratory without proper supervision.** One of your best safety precautions is to have a knowledgeable person present who can spot potential hazards and handle an emergency should it arrive.

6. **Handle glassware carefully.** Glass can break and can cause nasty cuts.

7. **Learn the location of fire extinguishers, a fire blanket, first-aid kit, eyewashes, and safety showers in your lab**. Be sure you know how and when they are to be used.

For many of the experiments in this text, you will find specific safety notes highlighted in boxes. These must, of course, be followed. But you should develop a habit of being safety-conscious whenever you are in a laboratory.

Preparation and Properties of Gases in a Breath

Introduction

This experiment is designed to accompany Chapter 1 in *Chemistry in Context.* Because we began the textbook with taking a breath, we will begin the laboratory investigations by examining the most important gases in that breath, oxygen and carbon dioxide. (The most plentiful component of air, nitrogen, is not very reactive.) Oxygen (O_2) makes up 21% of the air in an inhaled breath and 16% of the air in an exhaled breath. Part of the oxygen you inhale combines with carbon compounds in your body to produce carbon dioxide (CO_2) and energy. The carbon dioxide is then exhaled. In this experiment, you will prepare samples of oxygen and carbon dioxide by means of chemical reactions, and then you will investigate some of their chemical properties.

Background Information

To prepare oxygen, you will decompose a familiar household product, hydrogen peroxide (H_2O_2). Because peroxide compounds are produced in living organisms but are toxic, most living organisms have developed enzymes (biological catalysts) that rapidly break down the peroxides. A **catalyst** is a substance that participates in a chemical reaction and influences its speed without undergoing permanent change. If you have used hydrogen peroxide to clean a cut or scrape, you have probably seen an example of the catalytic reaction. Hydrogen peroxide bubbles as it is applied to the cut because an enzyme in your blood causes the peroxide to decompose and produce oxygen gas. In this experiment, you will use potassium iodide as the catalyst to decompose hydrogen peroxide into water and oxygen.

$$2H_2O_2 \xrightarrow{\text{catalyst}} 2H_2O + O_2$$

Carbon dioxide will be prepared from another common household product, baking soda, which has the chemical name sodium bicarbonate ($NaHCO_3$). When hydrochloric acid (HCl) is mixed with sodium bicarbonate, they undergo a chemical reaction to form sodium chloride (NaCl), water (H_2O), and carbon dioxide (CO_2).

$$NaHCO_3 + HCl \longrightarrow NaCl + H_2O + CO_2$$

Both gases will be generated in "zipper" plastic bags, and you will have an opportunity to make observations about the reactions. Samples of the gases will be tested for flammability and for reactivity with a water solution of calcium hydroxide, $Ca(OH)_2$, also known as "limewater."

Finally, you will investigate what happens when these gases dissolve in water. In particular, you will determine whether or not they react with water to form an acid. (Acids are discussed in detail in Chapter 6 of *Chemistry in Context.*) To test for changes in acidity, you will use an acid-sensitive dye called an **indicator**. The indicator in this case, bromthymol blue, is blue in the absence of acid and yellow in the presence of acids. Thus, if a gas mixes with a water solution of bromthymol blue and the color changes from blue to yellow, this is evidence that the gas has reacted with the water to produce an acid. Air, oxygen, carbon dioxide, and exhaled air will be tested with bromthymol blue solution.

Overview of the Experiment

1. Prepare samples of carbon dioxide and oxygen in plastic "zipper" bags.
2. Test these gases (plus air and exhaled air) for reaction with limewater.
3. Test these gases (plus air and exhaled air) for acidity in water solution.
4. Test these gases (plus air and exhaled air) with a glowing splint.
5. Clean up.

Materials Needed

Chemicals	Equipment
• 2–3 grams of $NaHCO_3$ • 1/2 gram of KI • 10 mL of 10% H_2O_2 • 5 mL of 20% HCl • 2 mL bromthymol blue solution • 2 mL limewater, $Ca(OH)_2$	• 3 1-pint heavy-duty "zipper" style reclosable plastic bags • 10 plastic transfer pipets • 24-well wellplate • 2 wood splints (or toothpicks) • 2 plastic teaspoons • matches • about a 10-cm square of black paper

EXPERIMENTAL PROCEDURE

I. Generating the Gases

General Instructions: The gases will be generated in plastic zipper-style reclosable bags, and you need to be able to seal and unseal the bags quickly and easily. It is important to completely seal the bags. Before proceeding further, test to be sure that a sealed bag with air in it does not leak when you squeeze it gently.

A. Carbon Dioxide

1. Place a teaspoonful (about 2 grams) of sodium bicarbonate, $NaHCO_3$, in the bottom corner of a 1-pint heavy-duty "zipper" bag.

2. Fill a plastic transfer pipet with 20% hydrochloric acid, HCl. (**Caution: corrosive material.**) To fill the pipet, try to squeeze nearly all the air out and then let the bulb fill with liquid. A second squeeze may help in getting more of the air out.

3. Place the pipet with the hydrochloric acid into the plastic bag with the $NaHCO_3$. Smooth out the bag so it contains a minimum amount of air; then seal the bag. (Take care not to press against the pipet.)

4. Hold the sealed plastic bag as shown in *Figure 1.1* and slowly squeeze the pipet so that the acid drops onto the $NaHCO_3$. Observe carefully what happens (several changes should be apparent) and record your observations in the data table. **Keep the bag sealed.**

5. You should now have a sealed plastic bag partially filled with carbon dioxide (CO_2) gas. *Leave the bag standing upright (zipper on top) by leaning it against something on the lab bench.*

Figure 1.1 Sodium bicarbonate in the "zipper" bag

B. Oxygen

1. Place a small pinch of potassium iodide (0.5 gram or less) in the bottom corner of another 1-pint heavy-duty "zipper" bag.

2. Fill one or two plastic transfer pipets with 10% hydrogen peroxide (H_2O_2). **(Caution: corrosive material.)** See directions above for filling the pipets.

3. Place the pipet(s) with 10% hydrogen peroxide into the plastic bag with the yeast. Smooth out the bag so it contains a minimum amount of air and then seal the bag. (Take care not to press against the pipet.)

4. Hold the sealed plastic bag as shown in *Figure 1.1* and slowly squeeze one pipet so that the H_2O_2 drops onto the potassium iodide. If a second pipet was used, wait a few moments and then gently squeeze it also. (Try to get all of the liquid *out* of the pipets.) Observe what happens and record your observations in the data table. **Keep the bag sealed.**

5. You should now have a sealed plastic bag partially filled with oxygen (O_2) gas. *Leave the bag upright (zipper on top) by leaning it against something on the lab bench.*

C. Exhaled Air

1. Breathe in and then exhale into a 1 pint heavy-duty "zipper" bag, repeating this 2 or 3 times; then seal the bag when it is inflated with your breath.

II. Properties of Carbon Dioxide, Oxygen, and Exhaled Air

General instructions for the tests: Use a plastic transfer pipet-full of gas to perform each test on each of the gases you generated in this experiment. Use a fresh pipet (clean and dry) for each test. It helps to have two students working together to fill the pipets by the following method.

To get a pipet-full of a particular gas, squeeze the bulb to expel as much as possible of the air inside the pipet. Keep squeezing the bulb and slowly push the tip of the pipet against the zip seal at one corner of the plastic bag containing the gas to be tested. With a bit of practice, you will be able to just push the pipet tip so that the seal opens around it. Taking care not to touch the liquid or solid chemicals in the bag with the pipet, push the pipet tip into the bag. Then release the bulb so that gas enters the pipet. Quickly withdraw the pipet. As the tip leaves the bag, immediately reseal the bag along the "zip strip."

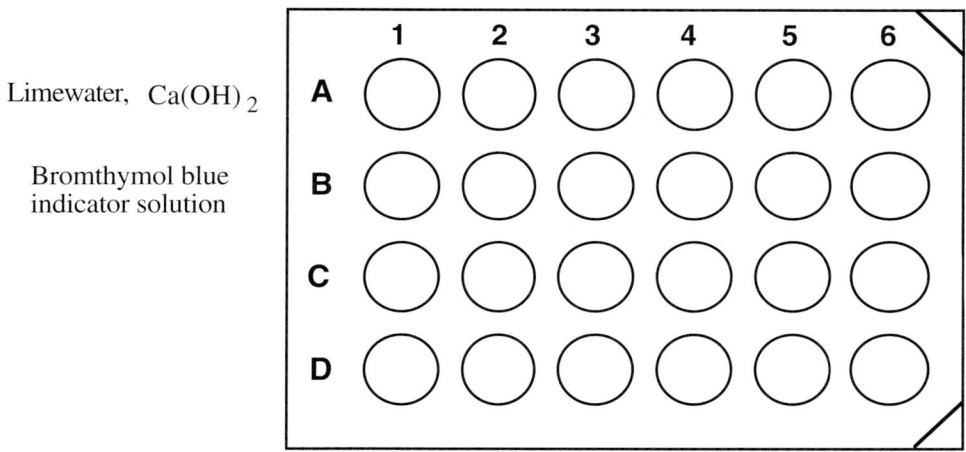

Limewater, $Ca(OH)_2$

Bromthymol blue indicator solution

Figure 1.2 The reaction wellplate setup

A. Limewater Tests

1. In row A of a wellplate (*Figure 1.2*), add 10 drops of limewater to wells A1, A2, A3, and A4.

2. Place the wellplate on a dark surface or a small piece of black paper.

3. Fill a clean pipet with carbon dioxide gas. Place the tip of the pipet into well A1. By gently squeezing the pipet, <u>slowly</u> bubble the gas through the limewater solution. Observe and record the results on the data sheet.

4. Repeat the test using a sample of oxygen gas bubbled into well A2. Record your observations.

5. Repeat the test using a sample of exhaled air bubbled into well A3. Record your observations.

6. Repeat the test using a sample of ordinary air bubbled into well A4. Record your observations.

B. Tests with an Indicator Solution

1. In row B of the wellplate, add 10 drops of bromthymol blue indicator solution to wells B1, B2, B3, and B4.

2. Fill a clean pipet with carbon dioxide gas. Place the tip of the pipet into well B1. By gently squeezing the pipet, <u>slowly</u> bubble the gas through the indicator solution. If the indicator changes color to yellow, it indicates that the solution has become acidic. Record the results on the data sheet.

3. Repeat the test using a sample of oxygen gas bubbled into well B2. Record your observations.

4. Repeat the test using a sample of exhaled air bubbled into well B3. Record your observations.

5. Repeat the test using a sample of ordinary air bubbled into well B4. Record your observations.

C. Glowing Wood-Splint Test

1. Use a clean, dry pipet to obtain another sample of carbon dioxide gas.

2. With a match, ignite the end of a wood splint or toothpick (or simply use a wood match). After it has burned for a few seconds, blow out the flame. Continue to blow on the embers so that they glow.

3. Have your lab partner hold the pipet filled with carbon dioxide so that the tip is very near the glowing ember and *gently* squeeze a puff of carbon dioxide gas directly at the glowing portion. Observe and record the results.

4. Using the same pipet, repeat this test using a sample of oxygen.

5. Finally, repeat this test using samples of exhaled air and ordinary air.

Clean Up

1. Rinse the plastic pipets with water.

2. Rinse the bags with water and discard the rinses down the drain. The bags can then be put in the trash basket provided. (Note: It is usually not a good idea to rinse laboratory chemicals down the drain. However, in this experiment, the chemicals are not very toxic nor are they considered environmental pollutants in the quantities used.)

Questions To Be Answered After Completing This Experiment
Write out answers to the following questions on a separate sheet of paper and hand it in along with the data sheet.

---

1. What role does the sodium bicarbonate play in the generation of carbon dioxide?

2. What role does potassium iodide play in the generation of oxygen? How does this differ from the role of sodium bicarbonate in the production of carbon dioxide?

3. Does exhaled air contain more CO_2 than normal air? What evidence do you have for your answer?

4. Based on what you observed about the interaction of carbon dioxide with the glowing splint, explain how CO_2 fire extinguishers work.

5. Based on what you observed about the interaction of oxygen with the glowing splint, explain why liquid oxygen is an extremely hazardous material.

6. Propose a way to estimate the volume of oxygen that was generated in the plastic bag.

7. **Optional challenge:** Propose a way to determine the amount of carbon dioxide in a carbonated beverage such as a club soda, beer, or cola drink.

Name _____   Date _____

Partner _____   Lab Section   _____

Data Sheet—Experiment 1

1. Observations regarding the generation of carbon dioxide.

2. Observations regarding the generation of oxygen.

3. Reactions with **limewater** [Ca(OH)$_2$ solution].

Gas	Observations
Carbon dioxide	
Oxygen	
Exhaled air	
Air	

4. Tests with **bromthymol blue acid-base indicator**.

Gas	Color of indicator	Is the solution acidic?
Carbon dioxide		
Oxygen		
Exhaled air		
Air		

Name _____   Date _____

Partner _____   Lab Section _____

Data Sheet—Experiment 1, page 2

5. Observations with a **glowing splint.**

Gas	Observations
Carbon dioxide	
Oxygen	
Exhaled air	
Air	

Chromatographic Study of Felt-Tip Pen Inks

Introduction

Chapter 1 in *Chemistry in Context,* fourth edition, introduces the concepts of pure substances and mixtures. This experiment utilizes paper chromatography to identify whether certain colored inks are pure substances or mixtures.

Chromatography is the study of separations of mixtures and is often used to identify unknown components in mixtures. By allowing the components to migrate along a flat sheet of paper (or, alternatively, through a long, narrow tube), they will move at different rates such that they become spread out and separated from each other. When this is done on paper, the usual strategy is to place small spots of substances near one edge of the paper, then place that edge in a liquid such as water and allow the liquid to move up the paper by capillary action. The liquid carries the components with it, but at differing rates depending on their solubility in water.

Background Information

The word "chromatography" derives from "color." Historically, the technique first utilized colors to see and identify the components of a mixture (much as you will do in this study). Today, a variety of other methods are employed to observe the separated components of mixtures. (Experiment 25 in this laboratory manual utilizes a slightly different chromatographic method to identify analgesic compounds.)

In this brief exploration, you will investigate several felt-tip pens of the type used with overhead projectors to find out whether each color is produced by a pure substance or a mixture of substances. By careful comparison, it should be possible to determine whether some of the pens have colored components in common.

Overview of the Experiment

1. Place spots of colored inks on a piece of filter paper.
2. Immerse one edge of filter paper in water and allow water to move the spots upward.
3. Compare the spot locations for the various inks.

Materials Needed

Equipment
• piece of filter paper, 7 cm x 12 cm
• ruler, metric scale
• 400-mL beaker
• glass rod
• set of colored pens (washable ink)

EXPERIMENTAL PROCEDURE

(This experiment should be done individually, so each person has a completed chromatogram, but you are encouraged to compare results and discuss the interpretation.)

1. Obtain four to six different-colored pens that contain vivid, washable inks. Good choices are the Vis-à-Vis overhead-projector pens or LiquidMark Washable Markers.

2. Obtain a rectangular piece of filter paper, approximately 7 cm x 12 cm. Write your name at the top (short edge). Using a *pencil*, draw a line 2 cm from the bottom edge and make *small* marks at 1-cm intervals along the line.

3. Place small dots of ink from several pens at the marks on the paper. Immediately make a record of which pens you used—preferably on the paper itself, *using a pencil, not an ink pen.*

4. Obtain a 400-mL beaker and glass rod to lay across the top of the beaker. Fold or roll the top edge of the filter paper so that it hangs on the glass rod inside the beaker but the bottom of the filter paper does not touch the bottom of the beaker.

5. Carefully add a small amount of water so that there is a shallow pool of pure water in the beaker with the bottom of the paper extending into it. Leave the beaker undisturbed on the desk and observe what happens.

6. When the top edge of the the water has moved about 2/3 of the way up the paper, remove it, and lay the paper out on a paper towel to dry. While still damp, <u>draw a pencil line to show the top edge of the wet area.</u>

7. **Report.** When the filter paper is completely dry, attach it to a sheet of 8 1/2 x 11 inch paper. Label the sheet with your name and lab section and include a brief summary of what samples and solvent were used. Answer the questions listed at the end of this experiment.

Optional Variations or Extensions

1. Investigate what happens with other kinds of pens. Some possibilities: other transparency pens, "Sharpie" or "Flair" pens, and ballpoint pens. (For ballpoint pens, make sure you have a good dark spot.) Use a fresh sheet of paper for each investigation. For ballpoint pens, try replacing the water with alcohol or acetone.

2. Investigate what happens if you replace the water with some other solvent, such as methyl alcohol, ethyl alcohol, or rubbing alcohol (which is a mixture of 70% isopropyl alcohol and 30% water).

3. Instead of placing the dots in a straight line near an edge of the paper, use a full circle of filter paper and place the dots along a small circle near the center of the paper. Place the paper horizontally over a petri dish (shallow glass dish) containing a small amount of water. Arrange a small paper cone in the center of the dish so that water is wicked up to the center of the filter-paper circle. Allow color development to occur. Observe and describe what you see.

Questions To Be Answered After Completing This Experiment

Write out answers to the following questions on a separate sheet of paper and hand it in along with the data sheet.

---

1. Propose one or more simple explanations for why the separations occurred. (There is more than one possible factor involved.)

2. Which of the transparency pen inks appear to contain a single colored substance, and which are mixtures?

3. Which pens have colored components in common with other pens? Explain your reasoning.

4. If you used a black pen, what colored components does it contain?

5. From these results, predict which colored components are most soluble in water and which are least soluble in water. Explain your reasoning.

6. (If you did optional step 2) Suggest a possible explanation for why the results using alcohol are different from those in water.

7. (If you did optional step 1) An interesting household problem is that of removing ballpoint pen stains from clothing. In light of your observations, can you suggest what will and will not likely work? What additional tests could you do to investigate this problem further?

Weighing Air and Cooling Water: A Graphic Experience

Introduction

Graphs provide an important and very useful way to present data in most disciplines. Graphs summarize the numerical data efficiently and are usually easier to understand and interpret than columns of numbers. Graphs are used frequently in *Chemistry in Context*, the text which accompanies this laboratory manual. In this assignment, you will collect some experimental data that lends itself to presentation in graphical form. You will then learn how to construct graphs in ways to make the visual presentations most effective.

Two short laboratory exercises are included. Your instructor will specify whether you are to do one or both of them; this will depend on the available time. The first exercise is a study of the relationship between pressure and mass of air in an empty soft-drink bottle. The second will investigate the relationship between time and temperature as a hot liquid cools.

Background Information on Constructing and Interpreting Graphs

Graphs are essentially a pictorial way of presenting information. For example, *Figure 2.1* presents some data on the relationship between mass and volume for a certain substance. The data is presented in both tabular and graphical form. It can be seen that there is a direct relationship between the mass and the volume for this substance.

Data Table

Volume	Mass
2 mL	1.0 g
4 mL	2.0 g
6 mL	3.0 g
8 mL	4.0 g
10 mL	5.0 g

Figure 2.1 Graphical representation of data. Volume as a function of mass.

The graph in *Figure 2.1* shows that as the volume increases, the mass increases in direct proportion. Data that produces a straight line when plotted is said to have a linear relationship. Representing data with a graph makes it very easy to estimate the value of the mass for a volume that is in-between measured data points, a process known as **interpolation.** As an example, if the volume were 5 mL, it is easy to see that the corresponding mass would be 2.5 g. It is also easy to extend the line of the graph and obtain data beyond the range of measured points, a process known as **extrapolation.** Thus, for example, if the volume were 15 mL, you should be able to convince yourself that the mass would be about 7.5 g.

To make graphs easy to read and interpret, the following conventions should be observed.

1. Graphs should be neat, legible, and well organized.

2. There should be a descriptive title designed to tell the reader what has been plotted. An example might be: "volume of plastic as a function of its mass." Titles such as "M vs. V" are too cryptic and should not be used.

3. The horizontal axis (x axis) and the vertical axis (y axis) should be clearly labeled to show what is plotted (e.g., volume) and the units (e.g., mL).

4. The data points should be plotted as dots with small circles drawn around them so that they show up clearly.

5. The *scale* of the graph should be chosen so that the graph fills as much of the paper as practical. In general, the scales on the x axis and the y axis do not need to start at 0. *Figures 2.1* and *2.3* show examples of correctly scaled graphs. The data points fill the graph space quite effectively. *Figure 2.2* shows three examples of incorrectly scaled graphs.

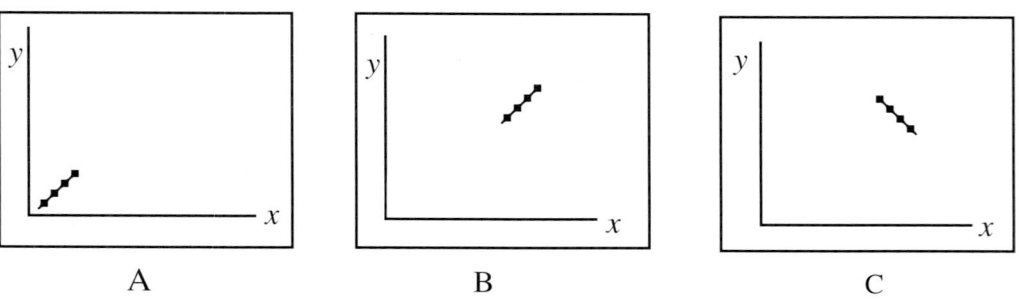

Figure 2.2 Examples of incorrectly scaled graphs

Graph A: The increments on the x and y axis should be made smaller to spread the graph out over the page.

Graph B: The starting points on the x and y axis should be changed and the increments on both the x and y axis should be made smaller so the graph takes up most of the page.

Graph C: The starting point on the x axis should be changed and the increments on both the x and y axis should be made smaller so the graph takes up most of the page.

6. If the data points appear to represent a linear (straight-line) relationship, use a ruler to draw a single straight line that best represents the average relationship. On the other hand, if the data points seem to follow a curve rather than a straight line, then draw the best *smooth* curve through them. (A curve is harder to draw than a straight line.) It is unlikely that all of the points will fit exactly on one smooth line (either straight or curved); therefore, some judgment must be exercised in deciding on the "best" fit. But it is not good form in science to draw zig-zag "connect-the-dot" lines.

7. When some sets of experimental data are plotted, it becomes apparent that certain points seem "out of line" with all of the others. (Scientists often call these "outliers.") In this case, it is

quite likely that some error was made, either in the original measurement or in writing it down. (It is also possible that an error was made in placing the point on the graph—this is the first thing to check.) If a point really seems to be out of line with the others, it is appropriate to exclude it when deciding on the best line. Clearly, this requires some judgment.

A straight-line relationship such as the one in *Figure 2.1* can be summarized with a simple algebraic equation of the form $y = mx + b$. In this equation, x and y are the values for the two quantities being plotted (e.g., volume and mass), m is the slope of the plotted line, and b is the **y-intercept** (the value of y when the line crosses the y axis or, in other words, the point where $x = 0$.) The **slope** of a line is calculated by determining the difference between the y values for two data points and the difference between the x values for the same two data points. The equation for this is

$$\text{slope} = m = \frac{y_2 - y_1}{x_2 - x_1}$$

The slope summarizes the relationship between the columns of data. In the example above, it is easy to see that the slope is 0.5, and it represents the relationship between mass and volume, or the density of the substance. Because the line crosses the y axis at 0, the y-intercept is 0, and the equation that represents the line is $y = 0.5x + 0$ or simply $y = 0.5x$. A straight-line relationship of the type shown in the figure is the easiest to construct and the easiest to interpret; therefore, researchers often go to great lengths to discover a linear relationship when a plot of raw data does not yield one directly.

Computer Aids to Graphing

Computer programs are available to take much of the guesswork and tedium out of constructing graphs. Many programs, for example, can create a complete graph from your data and provide you with a printed copy of the graph. Some programs can do what is called a "linear regression analysis," which uses a statistical method to calculate the best straight-line fit for a set of data points and then to draw the line. This can be particularly useful when there is a lot of "scatter" in the data, as in the example shown in *Figure 2.3*. Consult your instructor to determine if a computer-graphing option is available. Even if computer assistance is available, it is still useful to learn how to construct graphs on your own. And even with computer programs, some judgment is needed about choice of scales and whether any "outlier" points should be excluded.

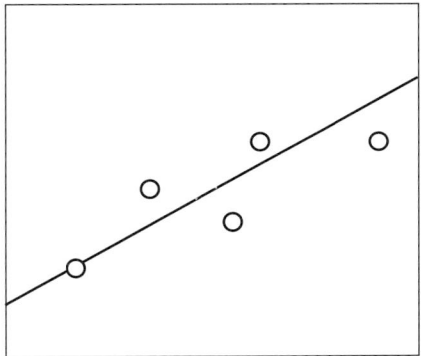

Figure 2.3 Linear regression fit of data points

Overview of the Experiment

PART I

1. Obtain an empty 2-liter bottle with a tire valve mounted in the cap.
2. Inflate the bottle to about 40 pounds per square inch with a bicycle pump.
3. Weigh the bottle and measure the pressure in the bottle.
4. Release some of the air; measure the pressure and weigh the bottle again.
5. Repeat step 4 several times until all of the air pressure has been released.

PART II

1. Heat some water in a beaker to a temperature of about 80°C.
2. Allow the water to cool and record its temperature at regular time intervals.

Materials Needed

- tire gauge
- hand bicycle pump
- plastic soda bottle (2 liter or 1 liter)
- bottle cap with a tire valve in it
- 250-mL beaker
- thermometer, 110°C

SAFETY NOTES

- It is important to wear safety glasses at all times in a laboratory. This is especially important for both parts of this experiment.

- Part II involves an open flame. No flammable chemicals should be in the vicinity. Long hair should be tied back, and extremely loose sleeves on clothing should be avoided.

EXPERIMENTAL PROCEDURE

I. Weighing Air

1. Before beginning the measurements, you need to learn how to use a laboratory balance for measuring the mass of an object. Your instructor will explain the use of the particular balances in your laboratory. If you are using an electronic-type balance, it is important to make sure the balance reads zero when empty (usually accomplished by pressing a TARE button). It is a good idea to recheck the zero each time you use the balance. For this experiment, you will need to make measurements to the nearest 1/100 of a gram (0.01 gram).

2. Obtain a clean, dry, plastic soda bottle (2-liter size preferred) with a screw cap that has been fitted with a tire valve. Screw the cap tightly onto the bottle.

3. Become familiar with using the tire gauge to measure the air pressure. You need to be able to push the gauge on squarely without letting out much air, and you need to know how to read the pressure units on the protruding stem. The pressure scale is probably marked in pounds-per-square-inch ("psi").

4. Attach the hose from a hand tire pump to the valve. Place the bottle in a protective enclosure (such as a wastebasket). Pump up the pressure in the bottle to about 40 psi, as measured by a tire gauge.

5. Weigh the bottle and record the mass. Wait 5 minutes and reweigh the bottle.

6. If the mass of the bottle (plus air) is the same as it was initially, the cap is sealed tightly and the experiment can be started. If the bottle's mass decreased by more than 0.05 gram, tighten the cap and repeat steps 4 and 5.

7. If necessary, pump up the pressure again to 30–40 psi. Carefully measure the air pressure in the bottle using the tire gauge and immediately record this in the data table. Then weigh the bottle on a balance and record the mass (to the nearest 0.01 gram).

8. Let a small amount of air out of the bottle. Measure and record the pressure again; then re-weigh and record the mass.

9. Repeat step 8 until you have obtained at least five sets of pressures and masses between your highest reading and 10 psi.

10. Finally, let all of the air out (so that the pressure is 0) and reweigh. (Of course the bottle is not really empty. It now has the same air pressure as the surrounding air, about 14.7 psi.)

11. **OPTIONAL:** These measurements can be done very rapidly. If time permits and you are interested, you might like to try repeating the study. (In scientific studies, the measurement technique often improves with repeated experiments.)

II. Cooling Water

1. Use a thermometer to measure the air temperature in the room. Record this on the data sheet.

2. Place a 100-mL beaker containing a thermometer and 50 mL of water on a ring stand over a Bunsen burner.

3. Light the Bunsen burner and heat the water to about 80°C. Turn the Bunsen burner off.

4. Without stirring or removing the beaker from the ring stand, allow the water to cool to 75°C.

5. When the water has cooled to 75°C, start recording the temperature at intervals of 2 minutes. Use a wall clock, wrist watch, or stop watch to keep track of the time. Record the temperatures to the nearest 0.5 degree.

6. Continue making measurements until the water has cooled to about 35°C.

III. Constructing Graphs of Your Data

I. Weighing Air

1. Look carefully at the range of recorded masses from lowest to highest. For the vertical (y) axis on the graph, select a *convenient* set of units that will include this range of masses. (It helps to select units with subdivisions of 10 since the masses are in decimal units.)

2. Similarly, look at the recorded pressures. (They should range from 0 to about 30 or 40 psi.) Choose a convenient set of units for the horizontal (x) axis.

3. Carefully plot the experimental points. (Use a pencil because it is easy to make mistakes in this kind of plotting.) Make small dots, and then draw small circles around them so that they show up clearly.

4. Examine the data points carefully. Using a ruler or other straight edge (preferably transparent), draw the "best" straight line through the data points. Don't do this too hurriedly—it requires some judgment. If the points are somewhat scattered, there should be equal numbers of them on either side of the line.

5. An interesting further use of this data set is to extrapolate the line to –14.7 psi. (If a computer-graphing program is available, this becomes easy to do simply by changing the scale on the horizontal axis.)

6. If specified by your instructor, calculate the slope of the line. To do this accurately, DO **NOT** use particular data points. Instead, select two places on the line near the opposite ends of the line. Carefully read off the x and y values for each, and then use the equation given in the Background section to calculate the slope.

II. Cooling Water

1. Select a convenient set of units on the horizontal (x) axis, starting at zero, to cover the number of minutes over which you made measurements.

2. Select a convenient set of units on the vertical (y) axis to cover the temperature range 35 to 75°C.

3. Proceed as in Part I to plot the points. Use pencil. Make small dots and draw a small circle around each one.

4. It is unlikely that these values will exhibit a straight line. Draw the best curve you can through the points. (Your instructor may demonstrate some tricks for drawing a smooth curve.)

1. For the graph of mass versus pressure, which do you think was more reliable, the mass or the pressure? Explain your answer. Make a general estimate of the probable uncertainties in each of your mass and pressure measurements. (To do this, think about the measuring device and how you used it.)

2. Looking at the graph for Part I, and taking into account your answer to Question 1, are you satisfied that there was actually a straight-line relationship between mass and pressure? Explain briefly.

3. Suppose you could pump some air out of the bottle rather than pumping in. How do you think the mass would change? Is there a limit to the change?

4. All of your mass measurements were actually the combined mass of the bottle itself plus the air it contained. How could you find out the mass of the air alone? (Hint: What is the significance of the weight of the bottle at −14.7 psi?)

5. Suggest a possible explanation for why the hot water did not cool down at a uniform rate and thus gave a curved line.

6. Approximately how long do you think it would take for the water to cool completely to room temperature? Explain your answer.

Name _____   Date _____

Partner _____   Lab Section _____

Data Sheet—Experiment 3

Part 1: Weighing Air
Initial mass of pressurized bottle _____   Mass after 5 minutes _____

Pressure (lbs/ sq. inch)	Mass of bottle

Part II: Cooling Water   Room Temperature _____°C

Clock time	Elapsed time	Temperature
	0.0 min.	

Name _____   Date _____

Partner _____   Lab Section _____

Graphs—Experiment 3

Solids in Cigarette Smoke

Introduction

Chapter 1 in *Chemistry in Context,* fourth edition, deals with air quality and various air pollutants. One of the important contributors to indoor air pollution is cigarette smoke. Cigarette smoking is a widespread habit in our society, and it is likely that the social, political, aesthetic, and medical issues resulting from cigarette smoking have affected smokers and non-smokers alike. The addictive nature of nicotine and the impact of cigarette smoking on the health of smokers has been extensively studied over many decades. In more recent years, the effect of so-called "second-hand smoke" or sidestream smoke, inhaled by nonsmokers has been evaluated and is now recognized as a significant issue for indoor air pollution. This laboratory investigation focuses on studying the smoke inhaled by smokers as well as sidestream smoke. You will use a simple apparatus to simulate the smoking process and will collect, weigh, and observe the solid particles produced by a burning cigarette. This will allow you better insight into the quantity and nature of the solids in cigarette smoke.

Background Information

A burning cigarette produces harmless carbon dioxide (CO_2) and water, but many of the other substances produced are not so harmless. Along with CO_2, carbon monoxide (CO) is produced. This toxic gas competes with oxygen in the bloodstream, reducing the amount of oxygen that is available to cells. Blood concentrations of carboxyhemoglobin, where the oxygen carrier hemoglobin is bound to CO instead of O_2, are about 10 times higher in a smoker than in a non-smoker. The nicotine in tobacco gets vaporized to a gas and is inhaled intact. Various minerals and carbon-based compounds are present as dust or ash. In total, cigarette smoke contains more than 4,000 different chemicals.

Some of the chemicals in cigarette smoke are proven carcinogens (cancer-causing substances). Others have the ability to serve as promoters of cancer when carcinogens are also present. According to the American Lung Association, cigarette smoking is the primary cause of lung cancer in the United States.

Emphysema is a chronic condition, affecting 2.0 million Americans living in the United States. It is a condition resulting from overinflation of structures in the lungs known as air sacs, or alveoli, which occurs when the alveoli walls begin to break down. The result is a loss of elasticity in the lungs as well as a decrease in the ability of the lungs to deliver oxygen to the bloodstream. In smokers, emphysema develops slowly after years of exposure to the irritation of cigarette smoke. Emphysema and chronic bronchitis together are classified as chronic obstructive pulmonary disease (COPD). Cigarette smoking is responsible for 83% of the cases of COPD.

Some of the compounds in cigarette smoke also affect nonsmokers who inhale so-called second-hand smoke. This is the smoke emanating from the burning end of the cigarette. It can be inhaled by the smoker as well as by nearby individuals and is also referred to as sidestream smoke. Inhalation of sidestream smoke is sometimes called passive smoking. In a 1993 report, the U.S. Environmental Protection Agency (EPA) concluded that 3,000 lung cancer deaths annually are due to passive smoking by nonsmokers.

Overview of the Experiment

1. Set up smoking apparatus or, if it is already set up for you, verify that connections are tight and that you understand how it works.
2. "Smoke" a filter cigarette to determine the amount of solids theoretically inhaled by the smoker and the amount of solids in second-hand smoke.
3. Repeat step 2 with another filter cigarette of the same brand.
4. Repeat steps 2 and 3 using a nonfilter cigarette.
5. Dispose of the filter paper, the burned cigarettes and ash, and clean up the apparatus.
6. Report your data and do the calculations.

Materials Needed

Equipment	Expendable items
• "smoking" apparatus (one per group of students) • ruler • balance	• filter cigarettes • nonfilter cigarettes • filter paper • matches

EXPERIMENTAL PROCEDURE

I. Setting Up the Apparatus

The apparatus is illustrated on the next page. Your instructor will probably already have it set up for you, but you should verify that you understand how it works and that the connections are tight. See the illustrations on the next page. Each apparatus is attached to a vacuum aspirator. This is a special adapter on the water faucet. When the water is turned on, the water flow draws air in from the side opening. This serves to pull air through the cigarette, allowing it to be "smoked." The second job of the vacuum aspirator is to pull the sidestream smoke through the appropriate filter. This allows you to measure the sidestream smoke AND prevents that smoke from entering the room and turning you into a second-hand smoker.

II. Smoking Cigarettes

Do a total of four trials. Carry out this procedure four times: **twice** with a filter cigarette and **twice** with a nonfilter cigarette. You may use the Camels provided, or you may test your own brand of filter or nonfilter cigarette if you obtain your instructor's permission. You need to do two trials on each type of cigarette you test.

1. Pretest the vacuum aspirator to make sure that when you turn on the water it drains into its sink and does not splash excessively. Your instructor will demonstrate approximately how fast the water flow should be. There may always be some splashing, so move notebooks and papers out of the way to keep them dry.

Cigarette

Glass funnel

Rubber tubing

Glass tubing

Rubber stopper

Filter paper

Rubber tubing to aspirator

Filter flask

2. Obtain two pieces of filter paper. On the outside edge of one, mark a "1" for first-hand or smoker smoke, and on the other, mark a "2" for second-hand smoke. Weigh each piece carefully using the analytical balance and record the weights.

3. Weigh a filter-tip cigarette. With a pencil, make a horizontal mark 5 cm from the end of the cigarette that will be lit. This is how far you will allow the cigarette to burn. Insert the end of the cigarette NOT to be lit into the funnel that will be attached to the "smoker's" flask.

4. Insert the two pieces of filter paper into the appropriate flasks and seal each flask with a rubber stopper with the tubing attached. It is important that you do not poke any holes in the filter paper, so the filter paper should be loose around the bottom of the stopper. However, there should be a good seal around the mouth of the flask.

5. Adjust the clamps holding the funnels so that there is about a 1/4 to 1/2 inch gap between the rims of the two funnels. A smaller gap will make it difficult for enough air to get in to burn the cigarette, and a larger gap will allow the sidestream smoke to escape into the room. You should keep the gap consistent for all tests.

6. Make sure your instructor checks your set-up before you light the cigarette.

7. Light the cigarette. At the same time, SLOWLY turn on the water to start the vacuum aspirator. At the appropriate flow rate, it should take about 3 to 5 minutes to smoke a cigarette down to the line. If your cigarette burns a lot faster than this, you should discard the data and repeat this part of the experiment.

8. When the cigarette has burned down to the line, break the connection between the smoker's flask and the aspirator and put out the cigarette (you don't have to collect the ashes on the cigarette). Leave the aspirator on 1 minute to finish collecting the second-hand smoke.

9. Break the connection to the second-hand smoke flask and then turn off the aspirator. Carefully remove the two pieces of filter paper and weigh each.

10. Weigh the unconsumed part of the cigarette (do not weigh the ashes).

11. Write down your observations about the appearance of each filter paper after the smoking exercise.

12. Throw the filter papers and the cigarettes in the trash, making sure the cigarettes are completely extinguished first. Wipe out the funnels with a paper towel to remove any solid residue. Reattach the flasks to the vacuum aspirator. Wash your hands.

13. Repeat the whole procedure with the other cigarettes until a total of 4 have been "smoked."

III. Calculations

A. Amount of Solids in Cigarette Smoke Relative to Total Mass Smoked

This calculation is to determine the amount of solids in the smoker's smoke and in sidestream smoke relative to the total mass burned when the cigarette was smoked. Carry out this calculation twice for each cigarette smoked: once for the smoker's smoke and once for the second-hand smoke.

First, calculate the mass of cigarette consumed and the mass of solids left on the filter papers:

mass of cigarette consumed = initial mass of cigarette – final mass after burning

mass of solids on filter paper = final mass of filter paper – initial mass of filter paper

Next, calculate the mass (in milligrams) of solid per gram of cigarette consumed:

$$\text{mg of solids per gram of cigarette consumed} = \frac{\text{mass of solids on filter paper}}{\text{mass of cigarette consumed}} \times 1{,}000$$

B. Solids in Second-Hand Smoke as a Percentage of Solids in Smoker's Smoke

This calculation gives a relative comparison of the amount of solids in second-hand smoke versus the amount that would theoretically be directly inhaled by the smoker. The higher this number, the greater the amount of solids in smoke that go into the room, rather than go into the smoker's lungs. This relative comparison makes it possible to compare one smoking experiment to another, since in each experiment, the actual amount of solids produced depends in part on how much of the cigarette was smoked.

$$\% \text{ second-hand smoke} = \frac{\text{mass of solids in second-hand smoke}}{\text{mass of solids in smoker's smoke}} \times 100$$

C. Average of Two Values

You should average the calculated values obtained in the two trials. It's important to do the calculations first and then average the results, rather than averaging the mass measurements and then doing the calculations.

$$\text{average} = (\text{value } 1 + \text{value } 2) \div 2$$

Sharing Results

Your instructor may ask you to post your results or share them in some way with other students. By looking at a large set of data, it should be possible to see patterns more clearly and also to confirm whether your results are consistent with those from other students.

> ## Questions To Be Answered After Completing This Experiment
> Write out answers to the following questions on a separate sheet of paper and hand it in along with the data sheet.

1. How do filter and nonfilter cigarettes compare with regard to the amount of solids in the smoke? Do filters make a significant difference? Do your results agree with what you would expect? Explain your reasoning.

2. What do you conclude from your data about the importance of second-hand smoke?

3. When repeat trials are run in any scientific experiment, it is important to see how well they agree with each other. Looking at your own data, do you see any significant discrepancies between the two trials for either type of cigarette? If so, you need to suggest a possible reason or reasons for the variation.

4. If class results were assembled, how well do your results agree with the rest of the class? If there is a significant difference, you need to suggest a possible reason or reasons for the variation.

5. When a cigarette is actually smoked, it is puffed or dragged on. In other words, it is not continually drawn on as are the cigarettes in this experiment. Tell how you could modify the apparatus and experimental procedure to more nearly mimic actual smoking.

6. Does the "second-hand smoke" data collected in this experiment accurately mimic the exposure of an actual passive smoker to the smoke of a single cigarette? To a roomful of burning cigarettes? Explain your answers.

Name _____   Date _____

Partner _____   Lab Section _____

Data Sheet—Experiment 4

Filter Cigarette (what brand?): _____

	Trial 1	Trial 2
Mass of cigarette consumed		
Initial mass of cigarette (g)		
Mass after burning (g)		
Mass of cigarette consumed (g)		
Smoker's flask (#1)		
Final mass of filter paper (g)		
Initial mass of filter pager (g)		
Mass of solids on filter paper (g)		
Milligrams of solids per gram of cigarette consumed (mg)		
Second-hand smoke flask (#2)		
Final mass of filter paper (g)		
Initial mass of filter paper (g)		
Mass of solids on filter paper (g)		
Milligrams of solids per gram of cigarette consumed (mg)		

	Trial 1	Trial 2
% of smoke as second-hand smoke		

Averages

Smoker's smoke: mg solids per gram of cigarette consumed _____ mg

Second-hand smoke: mg solids per gram of cigarette consumed _____ mg

Percent of smoke as second-hand smoke _____ %

Name _____ Date _____

Partner _____ Lab Section _____

Data Sheet—Experiment 4, page 2

Nonfilter Cigarette (what brand?): _____

	Trial 1	**Trial 2**
Mass of cigarette consumed		
Initial mass of cigarette (g)		
Mass after burning (g)		
Mass of cigarette consumed (g)		
Smoker's flask (#1)		
Final mass of filter paper (g)		
Initial mass of filter paper (g)		
Mass of solids on filter paper (g)		
Milligrams of solids per gram of cigarette consumed (mg)		
Second-hand smoke flask (#2)		
Final mass of filter paper (g)		
Initial mass of filter paper (g)		
Mass of solids on filter paper (g)		
Milligrams of solids per gram of cigarette consumed (mg)		

	Trial 1	Trial 2
% of smoke as second-hand smoke		

Averages

Smoker's smoke: mg solids per gram of cigarette consumed _____ mg

Second-hand smoke: mg solids per gram of cigarette consumed _____ mg

Percent of smoke as second-hand smoke _____ %

What Protects Us from Ultraviolet Light?

Introduction

Ultraviolet light has sufficient energy to cause changes in DNA and thus skin cancers. The dangers of exposure to ultraviolet light are described in Chapter 2 of *Chemistry in Context*. In recent years, depletion of the ozone layer has allowed more ultraviolet light to reach us, resulting in more cases of skin cancer. Consequently, we have become increasingly aware of the need to protect ourselves from exposure to ultraviolet light.

One strategy to avoid exposure to ultraviolet light would be to avoid exposure to any type of light. Few among us would be willing to take this to the extreme of moving underground, but we've all moved under a shady tree or used a hat to block out light.

A wide variety of commercial lotions are available with sun protection factors (SPF) ranging from 2 to 60. These lotions contain organic molecules such as oxybenzone and octyl methoxycinnamate that absorb light specifically in the ultraviolet region of the spectrum. Similarly, we can purchase sunglasses whose lenses absorb either UV-A (320–400 nm) or UV-B (280–320 nm) light, or both. Throughout the years, sunbathers have tried any number of homespun methods to protect themselves: slathering on baby oil, spraying mists of water, or sunbathing behind a glass or plastic panel.

The Problem

Your assignment is to design and conduct an experiment to test the ability of a minimum of three materials to protect us from ultraviolet light. You will be given a handful of beads that change color upon exposure to UV light. The more UV light they are exposed to, the faster their color will change. Thus, by measuring how long it takes the beads to change color, you are measuring the amount of UV light reaching them. You will also be provided with a variety of protective materials to test. Your instructor will give you a list of what is available and may allow you to bring additional items from home.

As you conduct your experiment, remember to protect the beads from all sides. You will need to take your experiment outdoors, but since the beads are quite sensitive, direct sunlight is not required. In fact, working in a shady spot usually gives better results. The beads will quickly change back to their original color if you put them in a dark place such as your pocket. Good experimental design requires controls. You want to change only one variable at a time. You will also want to know how the beads respond when left totally unprotected. Multiple trials should be used so that average data can be reported.

Overview of the Experiment

1. Design your experiment using a minimum of three protective materials and with appropriate controls included.
2. Get instructor approval/suggestions for your procedure.
3. Conduct your experiment.
4. Write your report.

Materials Needed

> • UV-sensitive beads
> • stopwatch or watch with second hand
> • a variety of protective materials such as sunscreen lotions with a variety of SPF ratings, hand lotion, baby oil, water, plastic wrap, solid pieces of clear or translucent plastic, glass plates or watch glasses, cheesecloth or thin cotton fabric, sunglass lenses, etc.

EXPERIMENTAL PROCEDURE

1. With your partner or other team members, develop a plan for your experiment. On your own paper, outline the procedure that you intend to follow. Be sure to include (a) three protective materials, (b) descriptions of controls, and (c) multiple trials. Specify what you will measure, how and when you will do it.

2. Obtain instructor approval to proceed. (Your instructor may offer some suggestions or ask questions to help you refine your experimental plan.)

3. Conduct your experiment. Keep a careful written record of what you do, noting especially any changes from your original plan. Record your data in an appropriate table of your own devising.

4. Prepare a written report, in which you should summarize
   a. the procedure you followed
   b. your data in clearly labeled tables
   c. your conclusions regarding
      i. how well each material protected the beads from UV light
      ii. your controls and multiple trials
      iii. any experimental difficulties
      iv. proposals for future improvement in the experimental design

---

Questions To Be Answered After Completing This Experiment

Write out answers to the following questions on a separate sheet of paper and hand it in along with the written report.

---

1. Compare the energy of ultraviolet light photons to that of infrared and visible light photons. Explain why UV light is potentially more dangerous than visible or infrared light.

2. What is the purpose of an experimental control? Give one specific example of a control you used and what information it provided for you.

3. Why are multiple trials of the same procedure needed? Give one specific example of an undesirable result that could occur if multiple trials were not used.

4. Describe one specific change that you would make in your procedure if you were doing the experiment over. Explain how this change could give better or more interesting results.

5. Most sunscreen lotions and sunglasses claim to protect against UV-A and UV-B light. Why don't they mention UV-C light? Is it less dangerous than the other types of UV light? What happens to the UV-C? (See Section 2.7 and Figure 2.5 of *Chemistry in Context,* fourth edition.)

6. On the next page, spectra of several materials that were tested for their ability to give protection from UV light are shown. Each spectrum shows the wavelength range(s) at which the material absorbs light. If a material has a large absorbance at a particular wavelength, it will protect us from light with that wavelength. If absorbance is low, the material will not provide effective protection. (For example, when the absorbance is 0.5, 70% of the light is being absorbed.)

   a. <u>On each of the spectra</u>, label these types-of-light regions along the *x*-axis:
      UV-A light 320–400 nm
      UV-B light 280–320 nm
      UV-C light below 280 nm

   b. Examine the spectrum of glass. Which types of light (UV-A, UV-B, and/or UV-C) does glass absorb best? From what type of UV light does this glass fail to protect us? Explain.

   c. Now look at the spectrum of clear plastic wrap. What types of light does the plastic wrap absorb? Would plastic wrap be good for preventing sunburn or skin cancer? Explain.

   d. What types of light does the sunscreen absorb?

   e. Finally, look at the spectrum of cotton fabric. How is this different from all the other spectra? Does the fabric selectively absorb certain wavelengths? What is the fabric doing that is different?

Ultraviolet Absorption Spectra of Protective Materials

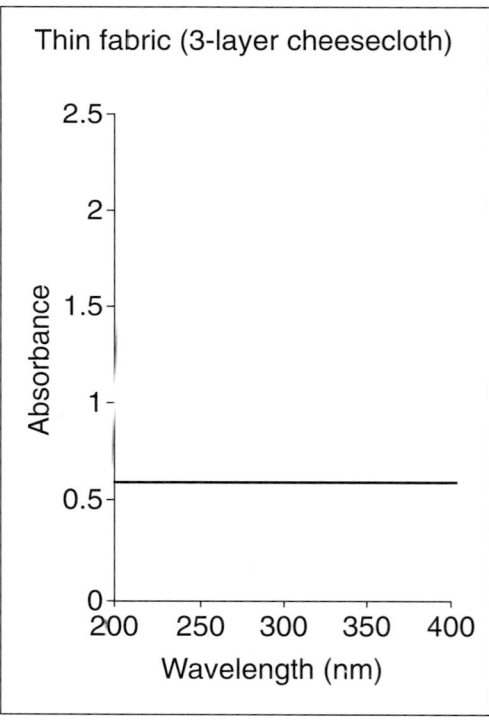

Visibly Delighted: A Spectrophotometric Study of Colored Solutions

Introduction

There is something magical about rainbows and colored light from glass prisms. The separation of white light into its colored components is the actual reason for these natural phenomena. In this experiment, you will investigate this partitioning of light in the visible region of the spectrum and its interaction with colored solutions. You will discover what colors correspond to what wavelengths, and, for each of several colored solutions, you will collect the necessary data to construct a spectrum showing how light is absorbed at different wavelengths for that substance. Measurements of this kind are widely used in nearly all areas of the physical and biological sciences.

An understanding of the electromagnetic spectrum and the interactions of radiation with matter is crucial to two of the case studies in *Chemistry in Context:* the stratospheric ozone shield (Chapter 2) and global warming (Chapter 3).

Background Information

The **electromagnetic spectrum** is discussed in Section 2.4 of the text. The region that the human eye can detect with the wonderful rainbow of colors (the so-called "visible" region) is only a small portion of the total electromagnetic spectrum. Each color and each unique position in the electromagnetic spectrum is identified by its **wavelength**. For the visible region, the lengths of the light waves are in the range of billionths of a meter, and, therefore, the wavelengths are expressed as **nanometers**, nm (1 nm = 1 x 10^{-9} meter).

The instrument used to measure the interaction of light with matter is a **spectrophotometer**. This device is designed to split visible light into its component colors (i.e., different wavelengths) and then allow light of a selected wavelength region to pass through a sample of the material being studied. An electronic detector measures the amount of light that has been transmitted or absorbed by the sample at each wavelength. All spectrophotometers have essentially the same major components but with varying degrees of sophistication. These essential components are (1) a light source, (2) a device to isolate or resolve particular wavelengths of light, (3) a sample holder, (4) a detector, and (5) a meter or other device to display the measured transmittance or absorbance of light. *Figure 6.1* (next page) shows a simple block diagram of spectrophotometer components.

Spectrophotometers are available for most regions of the electromagnetic spectrum. Spectro-photometers for the visible region of the spectrum were developed first and are still the most common. In these instruments, the light source is an ordinary incandescent lightbulb; the wavelength selector consists of a diffraction grating and some lenses; a liquid sample is placed in a test-tubelike container; and the detector is a phototube that converts light intensity into an electrical signal.

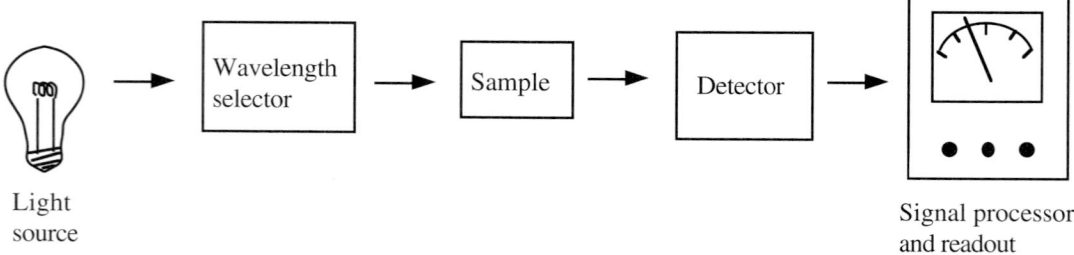

Figure 6.1 Block diagram of spectrophotometer components

Unfortunately, a simple spectrophotometer of this type can measure the transmittance of light of only a single wavelength at a time. Obtaining data over a range of wavelengths requires multiple measurements, each made at a different wavelength. Because light intensity and detector sensitivity vary with wavelength, each measurement must be corrected using a **blank**. For this experiment, the blank is simply a test tube containing pure water.

The meter on the instrument you use will display either **absorbance (A)** or **percent transmittance (%T)**. Absorbance is the more useful quantity because this is defined as the amount of light that passes through a sample of known thickness (the "path length") divided by the light that passes through a blank of equal thickness, and multiplied by 100.

$$\% \text{ transmittance } (\% \text{ T}) \; = \; \frac{\text{transmittance by sample}}{\text{transmittance by blank}} \; \text{ x } 100$$

In practice, this means you should use two matched test tubes or other containers: one containing the sample and one containing pure water. If the transmittance of the blank is first adjusted to read 100% at a particular wavelength, you can then read the percent transmittance directly from the meter on the spectrophotometer. From your transmission data, you will construct a graphical representation of the light transmittance of a colored compound over a range of wavelengths. This is accomplished by measuring the percent of the light transmitted at different wavelengths and plotting the wavelength on the *x* axis of a graph and the % transmittance on the *y* axis. The result is a **spectrum**, a curved line that shows how a particular solution transmits light in the visible region of the electromagnetic spectrum.

Many newer spectrophotometers display a different property, **absorbance**, instead of % transmittance. You may be using such an instrument for this experiment. In a general way, absorbance is the inverse of transmittance. Thus, for example, if a sample transmits all of the light at a certain wavelength, then the absorbance of light by the sample at that wavelength must be zero; conversely, if the transmittance is low, this means the absorbance is high. The mathematical relationship is slightly more complicated: Absorbance is defined as the negative logarithm of the transmittance, or $A = -\log T$. The particular utility of absorbance is that it is directly proportional to the concentration of the colored or absorbing material.

In this experiment, you will first determine the colors of different wavelengths of light throughout the visible region. By adjusting a wavelength dial, you can select particular wavelengths. Then you will determine which wavelengths of visible light are transmitted or absorbed by colored solutions. The colored liquids you will use for this part of the experiment are food-coloring dyes.

You will then investigate a substance of your own choosing. This can be any colored substance so long as it meets two conditions: (a) It must be transparent so that light will pass through it, and (b) it needs to have a fairly bright, distinct color.

One category of colored substances of particular interest to chemists is acid-base "indicators," colored dyes that are used to identify the acidic or alkaline character of solutions. You have probably already used one such substance (bromthymol blue) in Experiment 1. Others will be used in later experiments in this laboratory manual. Your instructor may ask you to do a spectrophotometric study of such an indicator, either in addition to, or in place of, the study of food-coloring dyes. If so, this will entail measuring the absorption spectrum of the indicator in both its acidic and alkaline forms.

Overview of the Experiment

1. Prepare a piece of chalk to reflect light from the spectrophotometer source.
2. Determine the color of light of selected wavelengths from 400 to 700 nm.
3. Measure the transmittance (or absorbance) spectrum for a red solution and a blue solution.
4. Optional: Also measure the spectrum for a green solution.
5. Plot a graph of transmittance (or absorbance) vs. wavelength for each colored solution.
6. Use this procedure to investigate a colored substance of your own choosing.

Materials Needed

Chemicals	Equipment
• half-inch long stick of chalk • solutions of food-color dyes • sample of a colored substance (provided by student)	• a simple spectrophotometer • 2–3 spectrophotometer test tubes • a holder for the test tubes

EXPERIMENTAL PROCEDURE

*[**Note**: The following instructions assume that you are using a Spectronic 20 Spectrophotometer, which is one of the most widely available simple spectrophotometers. If it is a digital version, your instructor will tell you whether to set it to read % T or absorbance. If you are using a different kind of spectrophotometer, your instructor will provide alternate instructions.]*

I. Looking at Light

1. Turn on the spectrophotometer and let it warm up. Investigate what the three control knobs do. (The knob on top is used to set the wavelength. The left front knob is the "zero" knob, used to adjust the meter to 0% T when the light path is blocked, and the right front knob is the "transmittance" knob, used to adjust the meter to 100% T when the blank is in place.)

2. Take a half-inch long piece of chalk and rub it on the blackboard until one end is worn down to a forty-five-degree angle (*Figure 6.2*). Place the chalk in the spectrophotometer test tube and insert it into the spectrophotometer so that the slanted side is pointing to the right (see diagram).

Worn-down chalk   Chalk in test tube

Figure 6.2 Chalk prepared for the spectrophotometer

3. Turn the wavelength dial so that the scale reads 500 nm (nanometers or 10^{-9} meter). Then turn the transmittance knob (front right) all the way to the right.

4. Look down the tube to the slanted piece of chalk. You should see a colored band of light. If not, raise, lower, or rotate the test tube slightly until you do. If you still cannot see the colored band, consult your instructor.

5. Now slowly turn the wavelength dial both directions and observe that the color of the light band changes.

6. If you turn the dial far enough in either direction, you will not see any color because you will be beyond the region that your eye can detect. By careful observation, find out the limits for your eyes and record these on the data sheet.

7. Finally, find out what wavelengths correspond to what colors. To do this, turn the wavelength knob so that the scale reads 400 nm. Look down the tube and record on the data sheet the color you see. Turn the knob so that the scale reads 440 or 450 nm and record the color you see in the data table. Repeat this at 40 or 50 nm intervals from 400 to 700 nm. You should have recorded a color corresponding to each wavelength you selected.

II. The Spectra of Red, Blue, and Green Solutions

1. Obtain at least two matched test tubes or other containers (sometimes called "sample cuvettes") that fit your spectrophotometer. Fill one of them about half full with a red dye solution. This will be the "sample." Fill the other tube half full with pure water. It will be the "blank." If you have a third tube available, fill it with a blue dye solution.

2. First adjust the "zero" on the spectrophotometer. With no tube in the sample holder and the *cover closed*, slowly turn the zero adjust knob (left front knob) until the meter needle or digital display shows exactly 0% transmittance. (For a meter with a needle, this requires placing your eye directly in front of the meter. If the meter has a mirror behind the needle, the mirror image of the needle should line up with the needle.)

3. Place the "blank" tube in the instrument and close the cover. Set the wavelength dial to 400 nm. Carefully adjust the right-hand front knob (100% T Adjust) until the meter reads exactly 100% T or 0 absorbance. (Again, for best results, your eye should be directly in front of the meter.)

4. Now remove the "blank" tube, insert the "sample" tube with the red solution, and close the cover. *Without changing any knobs*, carefully observe the meter reading and record this in the data table as % T (or absorbance) for the red solution at 400 nm.

     (**Optional:** You can check the accuracy of the reading by repeating steps 2–4. Small fluctuations in the electrical voltage can sometimes cause fluctuations in the meter readings. Repeated measurements can catch this, but they will also make the whole process much slower and may not be needed for this study.)

5. If you have a third tube with a blue solution, place it in the instrument, close the cover, then read and record the % T (or absorbance) in the data table.

6. Now change the wavelength dial to 420 or 425 nm. Repeat steps 3 and 4 by first inserting the "blank," closing the cover, and adjusting the meter to read 100% T (or 0 A). Then insert the "sample," observe the meter reading, and record the % T or absorbance. (Repeat with the blue solution if possible.)

7. Proceed in the same fashion, at 20- or 25-nm intervals, all the way to 700 nm. This sounds tedious, but in fact, it goes quite rapidly, especially if two students work together, one of whom records the data while the other switches tubes. (Remember that the 100% T (or 0 A) knob is adjusted *each* time you put the blank in, but you do not change *any* knob when the sample is inserted.)

8. Next, repeat the whole process with a blue solution if it was not done along with the red solution. Pour out the red solution, rinse the tube with water, and then fill it half way with a blue solution. Start at 400 nm and go up to 700 nm, recording the results in the data table.

9. You may either stop now and make a graph of these results (see Part VI) or you can simply examine the numbers to observe the pattern. In what wavelength region did this solution transmit the most light? In what wavelength region did it absorb the most light? Is this pattern consistent with the colors you observed in Part I?

10. Finally, obtain the spectrum of a green solution. First, make a <u>prediction</u> of what you think the spectrum will look like for a green solution, based on your results with the red and blue solutions. Show it to your instructor before proceeding. Then rinse and fill the sample tube with a green solution and proceed to record the transmittance or absorbance data for this solution in the same manner as for the other solutions.

III. Optional Extension or Alternate Assignment: Study of An Acid-Base Indicator

As noted in the Background section, acid-base indicators are colored dyes that change color in acidic and basic (alkaline) solutions. An interesting assignment is to measure the spectrum of such an indicator in both its acidic and basic forms—and even in its intermediate state halfway between the two forms.

If this assignment is done as an alternative to Part II, you can use the prepared table on the data sheet. If this is an additional study, you will need to construct a new table on a sheet of lined paper.

1. Rinse out the sample test tube and half-fill it with the blue indicator solution provided (bromthymol blue). A small amount of sodium hydroxide has been added to this solution in advance to be sure the solution is alkaline. (If a third test tube is available, rinse and half-fill it with the same indicator solution. Then add a drop of hydrochloric acid solution to make the solution acidic. It should change color to yellow.)

2. Proceed in the same manner as in Part II. At each wavelength chosen, first insert the "blank" test tube and set the 100% T; then measure the colored solution. Start at 400 nm and go up to 700 nm. If you have test tubes with both forms of the indicator, measure both of them before proceeding to the next wavelength.

3. If only one sample test tube was available, add a drop of hydrochloric acid to the colored dye, mix by gentle swirling, and then make measurements from 400 to 700 nm.

4. **Challenge:** Find a way to make an intermediate form of the indicator solution that is green. You will have to experiment by adding small amounts of very dilute acid (hydrochloric acid) and very dilute base (sodium hydroxide) to achieve a green color. Then measure the spectrum of this green solution and compare it with the spectra of the blue and yellow forms of the indicator.

IV. Optional: Investigation of Colored Substances of Your Own Choosing

There are many colored substances to choose from. Some common examples include beverages (such as Kool Aid, some soft drinks, and some juices), water solutions of jello, water-color paints, felt-tip pen inks, food color dyes, and hair dyes. You could even use a small piece of colored glass if it will fit in the test tube. As noted earlier, whatever you choose must be transparent (meaning that you can see through it), and it ought to have a fairly bright, distinct color.

You may have to experiment to find a suitable dilution of your substance so that the lowest transmittance does not drop below 5% T (or absorbance above 1.5). Then use the skills you have learned previously to collect data and plot a spectrum of the substance.

V. Optional: Ultraviolet or Infrared Spectra

Your instructor may demonstrate, or allow you to use, a spectrophotometer that measures ultraviolet or infrared spectra. You may be able to obtain an ultraviolet spectrum of a sunscreen material of the type used in Experiment 5. Or you may be able to obtain an infrared spectrum of one of the "greenhouse" gases such as carbon dioxide.

VI. Making Graphs of Your Results

Use the graph paper provided to plot the data for each colored solution or substance. The wavelength should be on the horizontal axis (*x* axis), ranging from 400 nm at the left edge to 700 nm at the right edge. Percent transmittance or absorbance should be on the vertical axis (*y* axis): % T should range from 0 at the bottom to 100 at the top. Absorbance should range from 0 at the bottom to 1.0 (or 2.0) at the top. First plot the red solution. *Using a pencil rather than a pen is preferable so that corrections can be made.* Carefully plot the data, making a small point with a circle around it for each measurement. When you are finished, draw a *smooth* curved line through the points as best you can. The line may not touch every point. (This is better than "connect the dots," which is likely to produce a somewhat jagged line.) Repeat this procedure with other colored substances. If you are careful, you can plot more than one spectrum on the same graph. Label each spectrum so that you know which is which. Your instructor will advise whether to plot the spectra separately.

Questions To Be Answered After Completing This Experiment

Write out answers to the following questions on a separate sheet of paper and hand it in along with the data sheet.

---

1. Answer these questions for the red solution.

   a. At what wavelengths does the solution have its greatest transmittance (lowest absorbance)? What color(s) of light corresponds to these wavelengths?

   b. At what wavelengths does the solution have its lowest transmittance (highest absorbance)? Here the dye molecules are absorbing the light. What color(s) of light is the red dye solution absorbing?

2. Answer these questions for the blue solution.

   a. At what wavelengths does the solution have its greatest transmittance (lowest absorbance)? What color(s) of light corresponds to these wavelengths?

   b. At what wavelengths does the solution have its lowest transmittance (highest absorbance)? Here the dye molecules are absorbing the light. What color(s) of light is the blue dye solution absorbing?

3. Review your answers to Questions 1 and 2, then propose a general rule relating the color of an object to the color(s) of light that it absorbs.

4. How does the energy content of light vary with wavelength (or color)? Consult Section 2.4 in the text for help. Which solution (red or blue) absorbed higher-energy light? Explain briefly.

5. Did the spectrum of a green solution look as you predicted it would? Explain briefly. If you did <u>not</u> investigate a green solution, predict what the spectrum should look like and explain your reasoning

6. If you did the optional study of an acid-base indicator, which form (blue or yellow) absorbs light of highest energy? Explain briefly.

7. If you studied a sample of your own choosing, describe the sample and the spectrum. Is the spectrum approximately what you predicted?

Name _____ Date _____

Partner _____ Lab Section _____

Data Sheet—Experiment 6

Wavelength limits for your eyes: lowest wavelength _____ highest wavelength _____

Wavelength (nm)	Color with chalk
400	

Wavelength nm	Red solution % T or Abs	Blue solution % T or Abs	Green solution % T or Abs	Student sample % T or Abs
400				

Name _____   Date _____

Partner _____   Lab Section _____

Graphs—Experiment 6

Wavelength (nm)

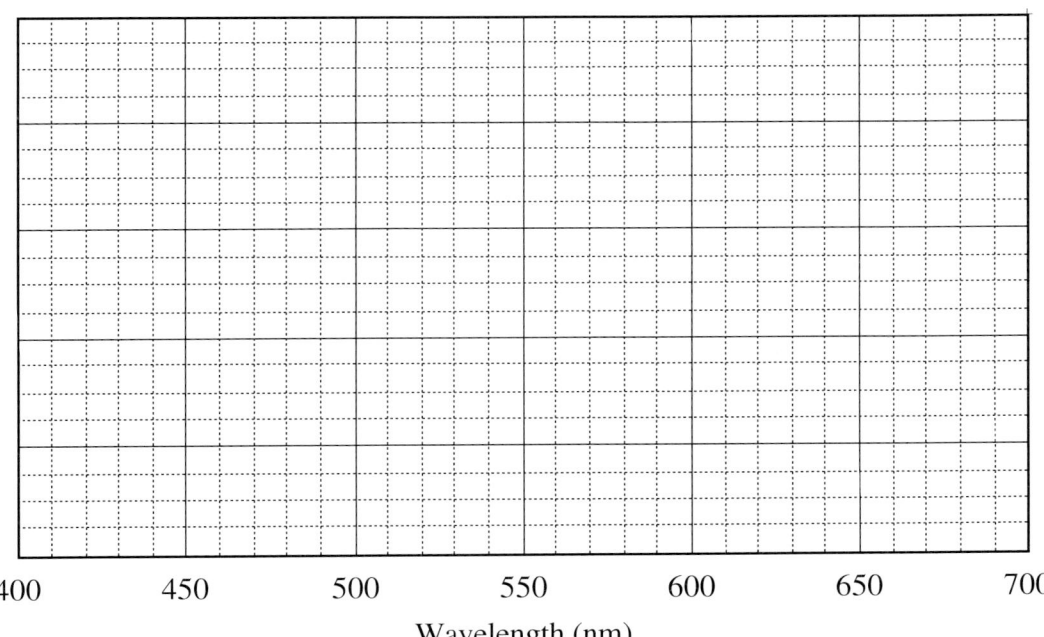

Wavelength (nm)

Name _____   Date _____

Partner _____   Lab Section _____

Graphs—Experiment 6, page 2

Wavelength (nm)

Wavelength (nm)

Chemical Bonds, Molecular Models, and Molecular Shapes

Introduction

The properties of chemical compounds are directly related to the ways in which atoms are bonded together into molecules. Chapter 2 in *Chemistry in Context* presents the basic ideas of chemical bonding, while Chapter 3 shows how the three-dimensional shapes of molecules are related to the bonding. In this exercise, you will have the opportunity to apply your knowledge from those two chapters by constructing simple ball-and-stick models for some common molecules. The models should help your understanding of electron arrangements in molecules and the resulting shapes of the molecules. You will investigate a number of small molecules containing carbon, nitrogen, oxygen, and hydrogen, as well as a few molecules containing fluorine, chlorine, or sulfur. These are mostly substances that are important in the atmosphere and in polluted air, as discussed in Chapters 1, 2, and 3 of *Chemistry in Context*. In the process of doing this exercise, you will see how "models" become very useful to chemists in understanding and predicting chemical properties.

Background Information

The existence of chemical compounds with fixed composition implies that the atoms in compounds must be connected in characteristic patterns. Early models showed the atoms hooked together like links on a chain. Modern representations are a good deal more abstract and often mathematical in nature. Nevertheless, it is possible to represent molecular structures with reasonable accuracy by using relatively simple models. The models serve as a three-dimensional representation of an abstract idea. Molecular model building has proven so useful that it is rare to find a chemist who does not have a model kit close at hand.

The chemical bonds that hold atoms together in molecules generally consist of *pairs of electrons* shared between two atoms. Atoms tend to share outer electrons in such a way that each atom in the union has a share in an *octet of electrons* in its outermost shell. This generalization has come to be known as the **octet rule**. (You should review the discussion of Lewis structures and the octet rule in Chapter 2 of the text.) The location of each element in the periodic table provides information about the number of electrons in the outermost level of the atoms. Carbon, for example, is in Group 4A and has four outer electrons; thus, it must share four additional electrons from other atoms in order to achieve a share in eight outer electrons (an octet). This is summarized in the table at the top of the next page. Oxygen, in Group 6A, has six outer electrons and shares two electrons from other atoms in order to achieve an octet. Hydrogen is a special case, needing to share its one electron with only one electron from another atom in order to achieve the stable outer electron configuration of the nonreactive element helium (He).

Electron Configurations in Atoms and Molecules

Atom	Outer electrons	Electrons needed from another atom	Electrons shared (no. of bonds)
Carbon	4	4	4
Nitrogen	5	3	3
Oxygen	6	2	2
Fluorine & chlorine	7	1	1
Hydrogen	1	1	1

A **single bond** consists of one shared pair of electrons; a **double bond** is two shared pairs (i.e., 4 electrons), and a **triple bond** is three shared pairs (6 electrons). On paper, the bonds are represented by single, double, or triple lines, respectively (–, =, ≡). In model kits, straight sticks represent single bonds, while double and triple bonds are represented by pairs or triplets of curved sticks or springs. Electrons not involved in bonding are termed *unshared electrons*.

An important part of this exercise involves identifying the *three-dimensional shapes* of molecules. (Molecular shapes are discussed in the text in Chapter 3.) Molecules have certain shapes depending on their component atoms and the ways in which they are bonded to each other. The important shapes encountered in this exercise are *linear, bent, triangular, pyramidal,* or *tetrahedral*. Several factors contribute to determining molecular shape. (1) Electron pairs (both shared and unshared) try to keep as far away from each other as possible, while still remaining "attached" to atoms. (After all, they are all negatively charged, and electrical charges of the same type will repel each other.) (2) Electron pairs tend to be symmetrically arranged around each atom in a three-dimensional manner. (3) Electron pairs *not* involved in the bonding ("unshared pairs" or "lone pairs") are equally as important as bonding electron pairs (shared pairs) in determining the overall molecular shape and arrangement of atoms.

Model Building Basics

Molecular model kits vary; therefore, your instructor will explain the particular models that you will use. The kit probably contains balls (used for atoms), sticks (used for single bonds and unshared electron pairs), and springs or curved sticks (used for double and triple bonds). Each stick or spring represents two electrons. Hydrogen atoms are usually represented by small, light-colored balls (yellow, white, or pale blue) that have only one hole. The color code for other atoms will vary. A common set of colors is shown in the table on the next page.

Note: There is one disadvantage to using the colored balls provided in most model sets. They usually have only enough holes for the correct number of bond pairs, and thus you will not be able to see the unshared electron pairs. An alternate approach is to use balls with four holes in them for all atoms other than hydrogen so that the octet (four pairs of electrons) will always be visible.

Typical Color Code for Molecular Model Sets

Atom	Color
Hydrogen	Yellow or white
Carbon	Black
Nitrogen	Blue
Oxygen	Red
Fluorine or chlorine	Green or purple
Sulfur	Yellow

1. Assemble the atoms required. (For example, to make the CH_4 molecule, you will need one black or blue sphere and four yellow ones.) Next, note the group in the periodic table to which each element belongs. The number of the group is also the number of outer electrons in an atom of the element.

2. To determine how many sticks (pairs of electrons) you will need, divide the total number of outer electrons by 2. For example, H_2O has one outer electron from each hydrogen and six from oxygen, for a total of eight. Hence, you will need four sticks to represent all the electrons in H_2O. Two sticks represent bonds between H and O, and two sticks represent unshared electron pairs.

3. If there is only one atom of one element in the molecule and more than one atom of another element, the single atom usually goes in the center of the molecule. This is the case in CO_2, but there are a few exceptions to this rule (such as N_2O, which has the arrangement NNO).

4. With the collected parts, assemble the model in such a way that each atom except hydrogen has a share in an octet of electrons. If you do not appear to have enough sticks (electron pairs) to give each atom (except hydrogen) an octet, try sharing more electrons by forming double or triple bonds (replace straight sticks with curved sticks or springs).

The Assignment

1. Each pair of students should have a model set. First, get acquainted with the components of the set. Note the holes in the various colored balls and their positions. If there are two lengths of sticks, the short ones are for bonds involving hydrogen, and the longer ones are for any other single bond.

2. Using the procedure outlined above, build models for each of the molecules listed on the data sheet. Then use information obtained from viewing the models to fill in the information in the last two columns. You should take time to think about (and write down in words and a diagram) the shape of each molecule before proceeding to the next one.

Questions To Be Answered After Completing This Experiment

Write out answers to the following questions on a separate sheet of paper and hand it in along with the data sheet.

---

1. The tetrahedral shape is one of the most fundamental shapes in chemical compounds. How would you describe it in words to someone who has never seen it?

2. The octet rule appears to be a very important rule governing the structures of molecules. In light of your work with models, provide a simple explanation for the importance of eight electrons.

3. Explain in your own words why nonbonded electron pairs help determine the shapes of molecules.

4. Do all of the assigned molecules obey the octet rule? If not, why (or in what way) did the octet rule fail?

5. As a test of what you have learned, predict the shapes of (a) NF_3, (b) H_2S, (c) Cl_2O.

6. Models do not necessarily have to be physical objects. They can be two-dimensional drawings or even mental constructs. Cite one or more examples of such models encountered outside of chemistry. Can you think of models that are used in your own field of study or that you will use in your future career?

Name _____   Date _____

Partner _____   Lab Section _____

Data Sheet—Experiment 7

Molecule	Total outer electrons	Lewis structure	Geometry of the atoms (Sketch the structural formula)
Methane CH_4			
Ammonia NH_3			
Water H_2O			
Oxygen O_2			
Nitrogen N_2			
Carbon dioxide CO_2			
Ozone O_3			

Name _____ Date _____

Partner _____ Lab Section _____

Data Sheet—Experiment 7, page 2

Molecule	Total outer electrons	Lewis structure	Geometry of the atoms (Sketch the structural formula)
CF_2Cl_2 (CFC-12)			
CHF_2Cl (CFC-22)			
Sulfur dioxide SO_2			
Carbon monoxide CO			
Formaldehyde H_2CO			
Optional Challenge Nitric Oxide NO			
Optional Challenge Nitrogen dioxide NO_2			

Weighing Gases To Find Molar Masses

Introduction

Chapter 3 of *Chemistry in Context* introduces the concept of **molar mass**, which is defined as the mass (in grams) that contains one Avogadro number (6.02×10^{23}) of atoms or molecules of an element or compound. However, the text does not explain in much detail how molar masses are determined. This experiment provides an opportunity to actually measure molar masses of several gases and then to compare your experimental results with the accepted values for those gases.

Background Information

The principle behind this experiment is of fundamental importance in chemistry. The key idea is that equal volumes of gases, at the same temperature and pressure, contain equal numbers of molecules. Therefore, if we compare the masses of equal volumes of two gases, we will also be comparing the masses of equal numbers of molecules of the substances. The *ratio* of the masses of equal number of molecules will be the same as the ratio of the masses of *individual molecules* as well as also being the ratio of the *molar masses* of the substances involved. Stated mathematically, this can be written as

$$\frac{\text{mass of } n \text{ molecules of gas A}}{\text{mass of } n \text{ molecules of gas B}} = \frac{\text{mass of 1 molecule of gas A}}{\text{mass of 1 molecule of gas B}} = \frac{\text{mass of 1 mole of gas A}}{\text{mass of 1 mole of gas B}}$$

In this experiment, you will weigh equal volumes of several different gases. You will then correct these data to yield the actual masses of the gas samples. Finally, you will use the relationship above to relate the corrected masses to a standard reference substance (oxygen, O_2) and thus obtain values for the molar masses of the gases that you have investigated.

The experimental procedure that you will use consists of weighing equal volumes of several gases in a container made out of a plastic bag. Because the actual mass of the gas samples is at most a gram or two, great care must be taken in weighing the sample. Weighing a gas turns out to be more complicated than weighing a small solid object. All objects are buoyed up by the air that surrounds them, and if the volume of the air displaced weighs more than the object displacing it, the object floats. The most common example is a helium-filled balloon. If a helium balloon were placed on a balance, it would appear to weigh nothing! Even if an object does not float, it is still affected by an upward force equal to the mass of air it displaces or proportional to the volume of air it displaces. You, for example, appear to weigh about 100 grams (1/4 lb) less in air then you would weigh in a vacuum, depending on your volume. Of course your mass, the amount of matter you contain, is the same in air or in a vacuum. In order to get an accurate weight for an object, the buoyancy effect must be corrected for by adding the mass of the air displaced by the object to the weight of the object in air. For small solid objects, the correction is so small it can be ignored in all but the most meticulous measurements. However, when weighing gases, this buoyancy effect becomes significant. The way to correct your data for buoyancy will be explained in the calculation section of this experiment.

Overview of the Experiment

1. Prepare a plastic bag to use as a gas container.
2. Fill the container with several different gas samples, including one unknown gas.
3. Weigh the container filled with each of the gases.
4. Determine the volume of the container.
5. Determine the mass of a liter of air at room temperature and pressure.
6. Calculate the apparent and corrected mass of each gas.
7. Calculate the molecular mass and volume for each gas.
8. From the data, propose a possible formula for the unknown gas.

Materials Needed

Equipment and supplies for each group	Equipment to be shared by two groups
• 1-quart plastic food bag • 1-hole rubber stopper with medicine dropper in the hole • large cork with center hole bored to fit the rubber stopper • several gases	• large empty glass or plastic bottle (2–5 liters), with a screw cap • large plastic pan if the sinks cannot be filled • 3–4 ft length of rubber tubing • 1–liter or larger (plastic) graduated cylinder

EXPERIMENTAL PROCEDURE

I. Weighing The Gases

Caution: A single drop of water will add enough weight to significantly distort the results. Therefore, check carefully to make sure the bag is dry at the start and do all of the weighing of gases before you attempt to determine the volume of the bag.

Note about weighing: The most critical measurements in this experiment are weights (of the gas container plus gases), which are obtained by means of a **laboratory balance**. Your instructor will explain how to use the particular balance(s) available in your lab. It is important to remember, first, that the balance must always read exactly zero when nothing is on the balance pan. (It is a good idea to check it before each weighing.) Second, the balance readings should be recorded to the nearest 1/100th of a gram (0.01 g), i.e., the second digit to the right of the decimal point. (If the last digit is zero, be sure to record that.) Some balances may show a third decimal place; if so, include it. On the other hand, with some balances, it may be difficult to read the second decimal place with confidence, in which case you will have to make the best estimate you can.

1. Obtain a plastic bag, a one-hole rubber stopper with a medicine dropper stuck through the hole, a cork ring sized such that the rubber stopper fits snugly in the ring, and a dropper cap (see *Figure 8.1* on the next page).

2. Push the opened end of the plastic bag through the center of the cork ring; then open the bag and push the rubber stopper firmly into the open bag so that the bag is tightly caught between the stopper and the ring. Make sure there are no leaks in the system. To do this, inflate the bag, put the rubber cap over the hole in the medicine dropper, and gently squeeze the bag. No air should escape.

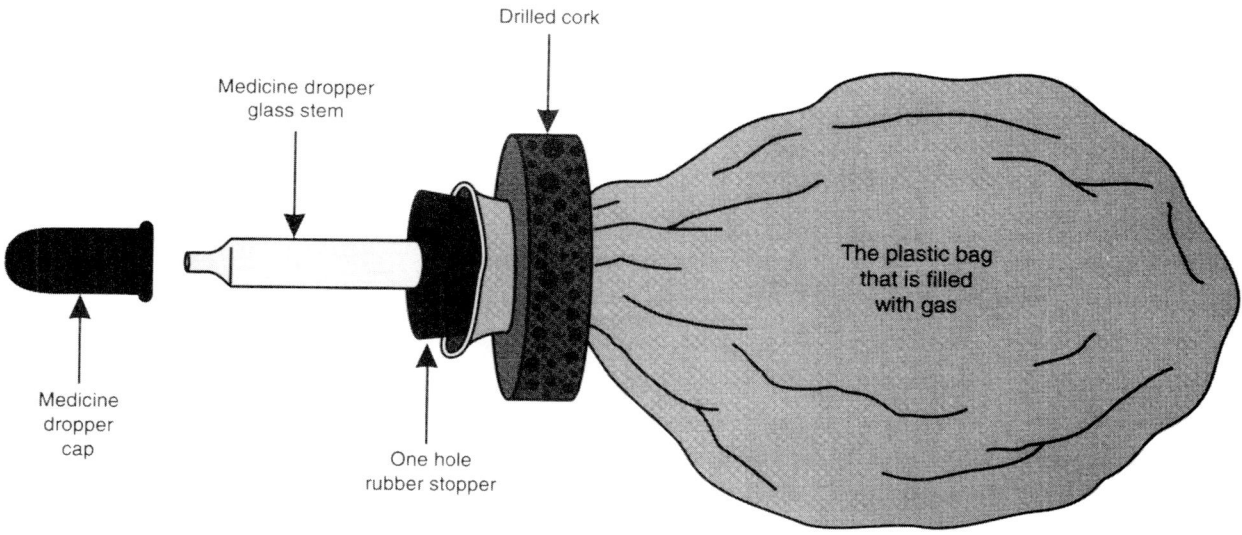

Figure 8.1 Diagram of the assembled system

3. With the rubber cap of the medicine dropper off, press out any air in the bag by smoothing it. It is important to get all the air out of the bag so that you weigh a completely empty bag.

4. Weigh the assembly, with the cap on the medicine dropper, to the nearest 0.01 gram and record the weight. (Remember to check first that the empty balance reads 0.00 g.)

5. Remove the cap from the medicine dropper and connect the bag via the medicine dropper to a source of gas.

6. Hold the bag by the stopper and completely fill the bag with one of the gases.

7. Allow the excess gas to escape so that the gas in the bag will be at room pressure but do not squeeze the bag. Replace the rubber cap.

8. Weigh the bag assembly containing the gas (which is at room temperature and normal atmospheric pressure) and record your result. Make sure that no part of the bag assembly hangs over the balance pan and touches any part of the balance frame or housing.

9. Completely empty the bag and repeat the procedure (steps 5–8) for the other gases available in the lab, including the unknown gas assigned to your team.

II. Finding the Volume of the Bag   (See *Figure 8.2*)

1. After you have weighed all the available gases, fill the bag with air at room pressure.

2. Attach a rubber hose to the bag at the medicine dropper.

3. Fill a large pan or the sink with water so that a large bottle of water can be inverted in it in such a way that the mouth of the bottle is below the surface.

4. Fill the large bottle with water (clear to the top, with no air space).

5. Cap the end of the bottle.

6. Invert the bottle so that the mouth is below the surface of the water in the sink or pan.   (The bottle should be completely filled with water; no air bubbles should be inside.)

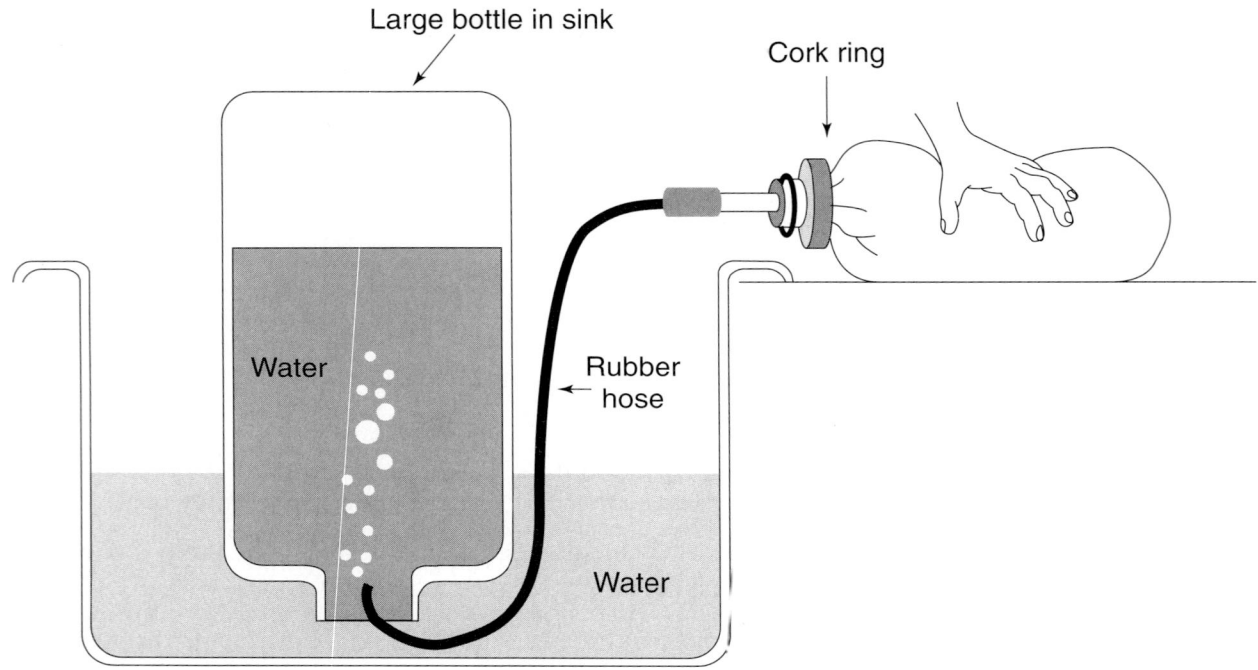

Figure 8.2 Diagram of the setup used to determine the volume of the plastic bag

7. Remove the cap from the bottle while the bottle opening is underwater.

8. With the help of your lab partner, insert the free end of the rubber hose into the mouth of the bottle, as shown in the diagram.

9. Gently squeeze the air out of the plastic bag. As you do so, bubbles of air will enter the bottle and force out some of the water in the bottle.

10. When the bag is empty, keep the bottle inverted with the mouth under water while you replace and tighten the screw cap. Then remove the bottle from the sink or pan.

11. Place the bottle in a normal upright position and remove the cap.

12. Obtain a 1,000-mL (1-liter) graduated cylinder. Fill it with water to exactly the 1,000-mL line. Slowly pour water from the cylinder into the open bottle until the bottle is completely filled. Then record the volume of water left in the graduated cylinder. Subtract the volume of water remaining in the graduated cylinder from 1,000-mL. This is the volume of water that was required to completely fill the bottle. It is also equal to the volume of air that was originally in the bag.

13. Record the room temperature and air pressure. Your instructor may provide these values for the entire class, or you may be asked to measure them with a thermometer and a barometer.

14. Use the chart on the back of the data sheet to determine the mass of 1 liter of air at the temperature and air pressure of the room. (**Note**: In the chart, air pressure is given in "mm Hg." This is the height of a column of liquid mercury in a mercury barometer. For comparison, standard atmospheric pressure at sea level is 760 mm Hg.)

III. Calculations (to be done for each gas sample)

1. Calculate the apparent mass of gas in the bag by subtracting the weight of the empty bag from the weight of the full bag. (This apparent mass may be less than zero. How can that be?)

2. Calculate the mass of air displaced by the bag of gas by multiplying the volume of the bag (in liters) by the mass of 1 liter of air at room temperature and pressure (taken from Table 8.1).

3. Calculate the corrected mass of the gas in the bag by adding the mass of the displaced air to the apparent mass of the gas.

4. Use the defined molar mass of oxygen (32 grams/mole) to calculate the experimental molar masses of the other gases, by means of the following equation.

$$\frac{\text{molar mass of gas A}}{\text{molar mass of oxygen}} = \frac{x \text{ g/mole}}{32 \text{ g/mole}} = \frac{\text{measured mass of gas A}}{\text{measured mass of oxygen}}$$

5. Record your results in the table on the data sheet.

> ## Questions To Be Answered After Completing This Experiment
> Write out answers to the following questions on a separate sheet of paper and hand it in along with the data sheet.

1. Compare your values for the molar mass of each gas with the accepted value. Do your results support the hypothesis that equal volumes of gases have equal numbers of molecules? Explain briefly the basis for your answer.

2. It is unlikely that the molar masses you determined exactly match the accepted values. What are some likely sources of error? Indicate whether each one would make your answer too low or too high. (**Hint:** How would the results of your experiment change if the temperature of one of the gas samples changed before it could be weighed?)

3. The unknown gas contains some combination of carbon, hydrogen, fluorine, chlorine, or sulfur. Not all elements are present in any one unknown gas. Suggest one or more possible molecular formulas for your unknown gas, assuming that the normal rules of chemical bonding for these elements are obeyed.

4. One of the first ways that was used to determine the molar mass of a liquid compound utilized the following procedure. The liquid was placed in a glass container, and the container was heated until all of the liquid had boiled away. As a consequence, the air originally in the container was totally displaced by gaseous molecules of the unknown compound. The container was sealed (while still hot), then cooled and weighed. What information do you think was needed to complete the determination of the molar mass of the liquid?

Name _____   Date _____

Partner _____   Lab Section _____

Data Sheet—Experiment 8

Room temperature (°C) _____   Atmospheric Pressure (mm Hg) _____

Weight of the empty bag assembly (in grams) _____

Final volume of water in the graduated cylinder _____

Volume of water added to the large bottle _____

Volume of the plastic bag when filled (in liters) _____

Weight of 1 liter of air at room temperature and pressure _____

Gas	Weight of bag assembly plus gas	Apparent mass of gas	Mass of air displaced	Corrected mass of gas	Calculated molar mass	Accepted molar mass
Oxygen					------	32
Burner gas (methane)						16
Carbon dioxide						44
Argon						40
Nitrogen						28
Unknown						

Possible formula for the unknown gas: _____

Table 8.1 The Mass (in grams) of 1 Liter of Air
at Different Temperatures and Pressures

Pressure	Temperature			
(mm Hg)	15°C	20°C	25°C	30°C
600	0.97	0.95	0.94	0.92
610	0.99	0.97	0.96	0.94
620	1.00	0.98	0.97	0.95
630	1.02	1.00	0.99	0.97
640	1.03	1.01	1.00	0.98
650	1.05	1.03	1.02	1.00
660	1.07	1.05	1.03	1.01
670	1.08	1.06	1.05	1.03
680	1.10	1.08	1.06	1.04
690	1.11	1.09	1.08	1.06
700	1.13	1.11	1.09	1.07
710	1.14	1.12	1.11	1.09
720	1.16	1.14	1.12	1.10
730	1.18	1.16	1.14	1.12
740	1.19	1.17	1.15	1.13
750	1.21	1.19	1.17	1.15
760	1.22	1.20	1.18	1.16
770	1.24	1.22	1.20	1.18

Chemical Moles:
Converting Baking Soda to Table Salt

Introduction

During chemical reactions, substances combine with each other in a definite proportion by mass, meaning that only a certain amount of one reactant will react with a given amount of another reactant. The amounts of reactant species can be expressed in a variety of ways: grams, pounds, tons, and liters. However, no matter what units are used, they are all related to the ratio of *moles* of one species reacting with a certain number of *moles* of another species. If you are unclear about the concept and definition of a **chemical mole**, you should review the discussion of moles in Chapter 3 of *Chemistry in Context*.

In this experiment, you will investigate in more detail a reaction you have already used in Experiment 1, i.e., the reaction of sodium bicarbonate, $NaHCO_3$, with hydrochloric acid, HCl. Sodium bicarbonate is also known as sodium hydrogen carbonate, but it is most familiar as baking soda. The reaction of sodium bicarbonate with hydrochloric acid produces salt (sodium chloride, NaCl), water, and carbon dioxide.

$$NaHCO_3 + HCl \longrightarrow NaCl + H_2O + CO_2$$

In Experiment 1, you used this reaction to make carbon dioxide in a plastic bag. Now you will do quantitative measurements to find out how many moles of salt (NaCl) are formed from 1 mole of $NaHCO_3$. You cannot measure moles directly, but you will determine the mass of $NaHCO_3$ used and the mass of NaCl formed by weighing samples on a balance. To further improve the accuracy of the experiment, data will be assembled from the whole class, and you will be asked to decide on a "best" value for the ratio.

Overview of the Experiment

1. Label and weigh three test tubes.
2. Add a weighed quantity of sodium bicarbonate to each test tube.
3. React the sodium bicarbonate with 10% hydrochloric acid.
4. Evaporate the liquid remaining in the test tube after the reaction takes place.
5. Determine the weight of sodium chloride produced.
6. Calculate the ratio of moles of NaCl formed to moles of $NaHCO_3$ used.

Materials Needed

Chemicals	Equipment
• sodium bicarbonate, $NaHCO_3$ • 10% hydrochloric acid, HCl	• 3 test tubes • test-tube holder • Bunsen burner

Use of a Laboratory Balance for Weighing

This experiment requires careful weighing of test tubes on a laboratory balance. Your instructor will explain how to use the particular balance(s) available in your lab. It is important to remember that the balance must always read exactly zero (0.00 g) when nothing is on the balance pan. It is a good idea to check it before each weighing.

EXPERIMENTAL PROCEDURE

SAFETY NOTES
- 10% HCl is corrosive to the skin and other materials. Avoid spilling it on yourself, your partner, or your work space.
- Be extremely cautious about flammable materials near an open flame.
- Be alert to the possibility of hot solution accidentally splashing out of a test tube if it is heated too rapidly.

1. Obtain three test tubes that are clean and completely dry. Label them A, B, and C, putting the labels near the tops of the tubes. Add a small boiling chip to each test tube.

2. Take the test tubes plus the lab manual or data table to one of the laboratory balances. Weigh each of the test tubes (including the boiling chip) to the nearest hundredth of a gram (0.01g) and record the masses in the data table, line 2. (If the second decimal place is a zero, be sure to record it, e.g., 18.10 g).

3. To each test tube add just enough $NaHCO_3$ to fill the curved bottom of the tube.

4. Again, weigh each test tube to the nearest hundredth of a gram (0.01 g) and record its mass (tube + contents) in the data table, line 1. **Note:** The masses of the three solid samples do not need to be identical. Typical masses of added $NaHCO_3$ will be about 0.3 to 0.7g.

5. Fill a plastic pipet with 10% hydrochloric acid solution. Add this acid dropwise to tube A. Let the liquid run down the wall of the test tube and gently "spank" the tube after each drop reaches the bottom. Continue to add acid slowly until all of the solid has dissolved. (It is important to add only the minimum amount of acid needed to get the solid dissolved.) Save the tube and its contents for further work.

6. Repeat step 5 with each of the remaining test tubes (B and C).

7. Gently heat test-tube A and its contents over a Bunsen burner flame, holding the tube at an angle and pointed away from you (as well as pointed away from anyone else in the immediate vicinity). See *Figure 9.1*. The idea is to evaporate the water in the tube without spattering anything out of the tube.

    Caution: Too rapid heating of the tube, especially if it is held in an upright position, will cause the hot contents to splash out of the tube and will necessitate starting over with a fresh sample. Boiling chips should help to produce smooth boiling. Your instructor may provide additional advice on how to minimize the problem.

Continue heating until *all* of the liquid has evaporated and solid NaCl remains. (It is crucial to the success of this experiment to be sure that all of the water has evaporated from the *upper* part of the tube.)

Figure 9.1 The correct way to heat a test tube over a Bunsen burner

8. Remove the tube from the flame and test for the evolution of water vapor from tube A by inverting a clean, dry test tube over the upright mouth of test-tube A. If condensation occurs in the cold test tube, continue the drying and testing process until no condensation occurs. Then set test-tube A with its dried contents aside to cool.

9. Repeat the procedures in steps 7 and 8 with test-tubes B and C.

10. Allow the three test tubes to cool (at least 5 minutes), *check to be sure there are no water droplets left*, and then weigh each with its contents. Record the masses in the data table.

11. If time permits, confirm that the tubes were fully dried by reheating them for 1–2 minutes, cooling for 5 minutes and reweighing. Record the second mass for each. If they were dry the first time, there should be a negligible change in mass.

Clean Up
Rinse the test tubes thoroughly with tap water, then with deionized water, and leave them upside down to drain.

Optional Extension or Alternate Assignment

After completing this study with sodium bicarbonate, you might wish to try a similar study with sodium carbonate (Na_2CO_3) to see how the results differ. Your instructor may ask some or all members of the class to do the experiment with sodium carbonate instead of sodium bicarbonate. The experimental procedure is the same, and the calculations are almost the same. If you do this version, you will need to make the appropriate changes in column 1 of the data table.

Calculations

1. For the three columns in the data table, subtract the numbers on line 2 from line 1 to obtain the masses of $NaHCO_3$ used (line 3).

2. Add up the molar masses of the atoms in $NaHCO_3$ to find the mass of 1 mole of $NaHCO_3$. Then use this result and the masses on line 3 to calculate the number of moles of $NaHCO_3$ used (line 4).

3. Line 6 (mass of tube) is the same as line 2. Subtract line 6 from line 5 to obtain the masses of $NaCl$ formed by the reaction (line 7).

4. Proceeding the same way as in step 2, calculate the moles of $NaCl$ formed (line 8).

5. Finally, calculate the <u>ratio</u> of moles $NaCl$ to moles $NaHCO_3$ (line 9). This should be recorded to two decimal places (i.e., 2 digits after the decimal point).

6. Calculate the average of your three experimental results for this ratio and record it to two decimal places.

Questions To Be Answered After Completing This Experiment

Write out answers to the following questions on a separate sheet of paper and hand it in along with the data sheet.

---

1. If class results have been assembled, what was the <u>average</u> of the mole ratios for the whole class? What was the <u>range</u> of experimental values (from lowest to highest)? Do you think any results should be excluded from calculating an average ratio for the whole class? If so, explain why and then calculate a revised average ratio.

2. Copy the balanced equation for the reaction of sodium bicarbonate with hydrochloric acid from the introduction to this experiment. What ratio for moles $NaCl$/moles $NaHCO_3$ does this balanced equation predict?

3. Do your experimental results and those from the class agree with the balanced equation? Write a brief discussion of the agreement (or lack of agreement).

4. Experimental error is a part of any scientific experiment. Suggest **two** possible sources of error in this experiment. (Do not include weighing errors.) Indicate whether each error would increase or decrease the experimental value for the ratio of moles $NaCl$ to moles $NaHCO_3$.

5. Suppose that you had used sodium carbonate, Na_2CO_3, instead of sodium bicarbonate. Balance the equation $Na_2CO_3 + HCl \longrightarrow NaCl + CO_2 + H_2O$. What mole ratio, moles $NaCl$ to moles Na_2CO_3, does this balanced equation predict?

Name _____   Date _____

Partner _____   Lab Section _____

Data Sheet—Experiment 9

	Tube A	Tube B	Tube C
1. Mass of tube + $NaHCO_3$			
2. Mass of tube			
3. Mass of $NaHCO_3$			
4. Moles of $NaHCO_3$			
5. Mass of tube + NaCl			
6. Mass of tube			
7. Mass of NaCl			
8. Moles of NaCl			
9. Ratio: $\dfrac{\text{moles of NaCl}}{\text{moles } NaHCO_3}$			

Average ratio for the three results: _____

Class average for the ratio: _____   (lowest: _____ highest: _____)

Revised class average (if applicable): _____

Hot Stuff: An Energy Conservation Problem

Introduction

This laboratory exercise is a departure from the usual experiment. It is an attempt to simulate the kind of problem solving that takes place in a scientific laboratory. There are no instructions, procedures, or data sheets. There is simply a problem to solve that requires reasoning skills and the application of previous experience or knowledge.

A student team will typically consist of three or four members, one of whom should be designated as the recorder for the group. This person records ideas, data, experimental procedures, etc. Your group might also want to assign other tasks to members of the group, for example, a "go-fer," an experimentalist, a report writer, a team leader. (Your investigation will be more efficient if you take a few minutes at the beginning to be sure each person has a clear role to play.) Your team must devise a method for solving the problem and then obtain a numerical answer. When finished, your instructor will lead a class discussion in which teams will compare answers and methods.

The Problem

As an energy conservation method, you decide to turn down your home water heater so that it only heats water to 55°C. (It previously heated the water to between 60°C and 70°C.) Unfortunately, when you decide to measure the temperature of the water in your water heater, the only thermometer available has a maximum temperature of 40°C. In the laboratory, there is a simulated water heater, consisting of a large coffee pot filled with hot water. Each team will have available a thermometer that has the graduations above 40°C covered so that you cannot see them. Using your 40°C thermometer and materials available to you in the lab, devise a way to measure the temperature of the water in the coffee pot. Your team should be prepared to defend your answer and the method you used to obtain it.

Optional: Your instructor may ask you to propose two or three *different* methods for solving this problem, then test each of them, decide which gives the most reliable answer, and explain why.

Materials and Equipment

A variety of materials and equipment will be available that you can use to solve this problem. At a minimum your "lab" should have graduated cylinders, burets, beakers, flasks, plasticware, Styrofoam cups, stirrers, test tubes, plastic pipets, a balance, and, of course, a 40°C thermometer.

Reporting Your Results

When everyone is finished, your class will assemble to hear a report from each student team about the method used and the results. Therefore, it is important for your team to keep a complete record of everything you do and the numerical data you obtain. In addition to an oral report to the class, your group should prepare a brief (one-page) written report of your investigation. Your instructor will specify what should be included in the report.

Comparison of the Energy Content of Fuels

Introduction

The ready availability of high-quality energy sources is one of the most important issues facing the global community in the late twentieth century. Chapter 4 in *Chemistry in Context* examines some of the societal choices we face and the advantages/disadvantages of various fossil fuels. Hydrocarbons (and mixtures of hydrocarbons) are among the most energy-rich and cleanest-burning fuels. Many of them are liquids, which make them attractive for transportation fuels.

In this experiment, you will investigate the energy content of several hydrocarbon fuels by using them to heat water. The data that you and your classmates obtain will enable you to compare fuels to see which ones provide more energy for a given mass of fuel burned. You will also be able to draw conclusions about the energy implications of using oxygenated fuels such as ethanol.

Background Information

Fuels are substances that burn to give off relatively large amounts of heat. In an overall sense, such burning is simply a combustion reaction between the fuel and oxygen. The amount of heat generated depends on what kind of fuel is used and how much of it is burned.

The simplest example of a common fuel is natural gas, which is almost pure methane (CH_4). When methane is burned completely, the only products are carbon dioxide and water.

$$CH_4 + 2\,O_2 \longrightarrow CO_2 + 2\,H_2O$$

Methane is an example of a hydrocarbon fuel. **Hydrocarbons** are compounds that contain only hydrogen and carbon. Other hydrocarbons, such as propane (C_3H_8) and butane (C_4H_{10}), will burn to produce the same products, carbon dioxide and water.

Some fuels such as ethanol (ethyl alcohol, C_2H_5OH) contain oxygen in addition to carbon and hydrogen. In effect, they are hydrocarbons that have already partially reacted with oxygen:

$$2\,C_2H_6 + O_2 \longrightarrow 2\,C_2H_5OH$$

In this experiment, you will measure the amount of heat given off by known amounts of several fuels. Your instructor will tell you exactly which fuels will be studied and how the class assignments will be made. Some of the fuels will be alcohols such as methanol (CH_3OH), ethanol (C_2H_5OH), isopropanol (C_3H_7OH), or butanol (C_4H_9OH). Some will be hydrocarbons, such as lamp oil or candle wax. Each of the latter is actually a mixture of hydrocarbons, but we will approximate their compositions as $C_{12}H_{26}$ (lamp oil) and $C_{40}H_{82}$ (candle wax).

The experimental procedure is to heat some water by burning a measured amount of a fuel sample. It takes exactly 1 calorie (cal) of heat to raise the temperature of 1 gram of liquid water by 1°C. (In Section 5.9 of the text, this is called the "specific heat.") Therefore, if you know the mass of water and how many degrees the temperature goes up, then the total amount of heat absorbed by the water can be calculated as follows:

$$\text{heat absorbed (calories)} = \text{mass of water (grams)} \times \text{temp. change (C)} \times 1.00 \text{ cal/g°C}$$
$$= m \times \Delta T \times 1.00 \text{ cal/g°C}$$

ΔT is a shorthand way of saying "change of temperature." In this equation, the last term (1.00 cal/g°C) is needed to make the units come out right. (You should convince yourself that combining and canceling the units on the right side will leave only calories.) Theoretically, the amount of heat liberated by the burning fuel should equal the heat absorbed by the water, but in practice, some of the heat will be lost to the surroundings.

Overview of the Experiment

1. Assemble the apparatus and obtain a burner containing a known fuel.
2. Add a measured volume of water to the can and then determine the mass of the water.
3. Weigh the burner.
4. Record the initial temperature of the water.
5. Light the burner and heat the water until the temperature increases about 20°C.
6. Record the highest temperature of the water.
7. Weigh the burner again, in order to find the mass of fuel used.
8. Repeat with two or more additional trials.
9. For each trial, calculate the amount of heat released per gram of fuel burned.

Materials Needed

Chemicals	Equipment
• 1 or more "alcohol" lamps containing liquid fuels (to be assigned by the instructor)	• 1 empty soda can without a top • 1 thermometer • 1 ring stand with iron ring attached • 1 glass rod • 100-mL graduated cylinder • package of matches • top-loading balance (shared)

Important Note About Use of Laboratory Balances:
The success of this experiment depends heavily on the accuracy of many mass measurements. All weighing will be done on a laboratory balance, which you will be sharing with other students. Your instructor will explain the use of the particular balances in your lab. For each weighing, it is important to know that the balance reads *exactly zero* with nothing on the balance pan. For some kinds of laboratory balances, the "zero" can be changed easily, and perhaps unintentionally, so your measurements might be in error. With a balance of that type, it is important to check the "zero" *each* time you make a weighing.

EXPERIMENTAL PROCEDURE

The following general procedure is to be used for each measurement. Your instructor will tell you which fuel or fuels you are to investigate. You may be asked to do one trial with each of three different fuels. Alternatively, you may be asked to do several trials with a single fuel to improve the level of your confidence about an average value for that particular fuel, in which case the results for the whole class will be assembled for comparison.

> **SAFETY NOTES**
> • You need to be aware constantly that you (and other students) are working with flammable solvents and open flames. Handle the fuel burners very carefully. Before starting the experiment, be sure you know where a fire extinguisher is located in your laboratory.
> • As with any experiment involving open flames, long hair should be tied back and extremely loose sleeves should be avoided.

1. Obtain a dry soda can with the top removed and two holes punched on opposite sides near the top. Slide a glass rod through the holes in the can so that it can be suspended from the ring attached to a ring stand as shown in the diagram on the next page.

2. Obtain a fuel burner, place it under the can, and adjust the height of the ring so that the bottom of the can is about 2 centimeters above the top of the wick. (*Figure 11.1.*)

3. Take the empty soda can plus your data sheet to a balance. Check to be sure that the empty balance reads 0.0 gram; then weigh the can and record the mass to the nearest 0.1 gram.

4. Using a graduated cylinder, add approximately 100 mL of water to the can. Reweigh the can plus water to the nearest 0.1 gram. By subtraction, calculate the mass of water in the can.

5. Put a thermometer in the can, stir the water for a few moments, and then measure the temperature of the water, trying to estimate to the nearest 0.1 degree Celsius. (If you are using a glass thermometer, it is probably marked at 1 degree intervals, so it is not possible to measure the temperature accurately to a tenth of a degree. Nevertheless, it is useful to make the best estimate that you can.)

6. Take the fuel burner plus your data sheet to a balance. If necessary, check to be sure that the empty balance reads zero. Weigh the burner and record its mass on the data sheet, reporting it to the nearest 0.01 gram (or the nearest 0.001 gram if the balance permits this).

> *Note that the can and water need only be weighed to the nearest 0.1 gram, while the burner must be weighed to the nearest 0.01 or 0.001 gram.*

Figure 11.1 Diagram of can and burner setup

7. Place the fuel burner under the soda can and light the burner. Observe the flame. If necessary, cautiously adjust the height of the can so that the top of the flame is just below the bottom of the can.

8. Stir the water occasionally and continue heating the water until the temperature has increased about 20 degrees Celsius; then extinguish the flame. Do the next two steps quickly.

9. Continue stirring the water gently until the temperature stops rising; then record the highest temperature, again estimating it to the nearest 0.1 degree. Calculate the temperature change by subtracting the initial temperature from the final (highest) temperature.

10. Take the burner and your data sheet to a balance. If necessary, check that the balance reads zero and then weigh the burner and record the mass to the nearest 0.01 gram (or 0.001 gram). Calculate the mass of fuel burned by subtracting the final weight of the burner from the original weight of the burner.

11. Before doing another measurement, take a few moments to discuss the procedure with your partner. Did you encounter any difficulties? Can you think of any desirable improvements in the procedure? Also do the calculations, as described in the next section. This will give you a better idea of what you are looking for in subsequent trials.

12. Now repeat the procedure, either making additional measurements with the same fuel or switching to different fuels, as directed. If the water in the can has cooled to nearly room temperature, you can use the same water sample. If not, it is a good idea to empty the can and add a fresh 100-mL portion of water. Note that you recorded the mass of the *dry* empty can before you started. This will not change, except for possible buildup of soot on the bottom; therefore, you need to measure only the mass of the can with water in it.

Calculations

To calculate the amount of heat liberated by 1 gram of burning fuel, you need the following items of information from each trial, all of which should be on the data sheet: (a) the mass of water that was heated, (b) the change in temperature of the water, and (c) the mass of fuel that was burned.

Since it takes 1.00 calorie of energy to raise the temperature of 1 gram of water 1 degree Celsius, the total heat absorbed by the water is equal to the temperature change × the grams of water × 1.00 cal/gram degree.

$$\text{Heat absorbed} = \text{temperature change} \times \text{grams of water} \times 1.00 \text{ cal/gm}°C$$

1. For each trial, calculate the total calories of heat absorbed by the water. This will be assumed to be the same as the amount of heat liberated by the burning fuel.

2. Then calculate the calories of heat per 1 gram of fuel burned.

You should do these calculations for each trial as soon as it is finished to see how the results are coming out.

Interpretation of Results

Because it was only possible for you to do a few trials, it is desirable to assemble a much larger body of data from your whole class or lab section so that more reliable comparisons of fuels can be made. Your instructor will tell you how to post or report your results for the rest of the class to see. You may be asked to calculate class averages for each fuel that was used. The following questions should help to focus your interpretation of the results.

Optional Extensions

1. **Investigate the effects of changes in procedure**. These might include (a) adding some nonflammable insulation around the can, (b) adding a cylindrical shield of aluminum foil around the burner, (c) using 200 mL of water and only a 10°C temperature rise (or 50 mL of water and a 40°C temperature change).

2. **Energy content of wood**. See Performance-Based Assessment Activity #3.

3. **Caloric content of nuts**. See Performance-Based Assessment Activity #4.

Questions To Be Answered After Completing This Experiment

Write out answers to the following questions on a separate sheet of paper and hand it in along with the data sheet.

---

1. Rank the fuels in order from least to most "heat per gram of fuel." Now identify each fuel in your ranking as either a hydrocarbon or an alcohol. What general rule can you propose about the heat content of hydrocarbons versus that of alcohols?

2. Which class of fuels, hydrocarbons or alcohols, contains oxygen? Based upon your answer to Question 1, how does the presence of oxygen in the structure of a fuel affect its heat content? Give a reasonable <u>explanation</u> for the difference.

3. Gasohol is a mixture of gasoline (hydrocarbons) and ethanol (an alcohol).

   a. How would the energy content per gram of gasohol compare to that of plain gasoline?

   b. How do you think this would affect a car's fuel efficiency (miles per gallon)?

   c. Considering that gasoline comes from petroleum, while ethanol can be made from corn, explain why environmentalists might promote the use of gasohol. Now explain why farmers might promote the use of gasohol. Finally, explain how you think someone concerned with feeding the poor might feel about increasing use of gasohol.

4. Carbohydrates have the general formula CH_2O, while fats typically have a general formula of about $C_{10}H_{19}O$. Which of these classes of foods has the greater oxygen content as a percentage of the molecule's mass? Since foods are simply fuels for the body, which of these will release more energy when metabolized in the body? Explain your reasoning. How does this conclusion compare to the relative calorie content of fats and carbohydrates? (If you are not sure how many dietary Calories are in a gram of fat or carbohydrate, see Section 11.8 in your textbook.)

5. Suppose you put 50 mL of water in the can instead of 100 mL and heated it 40 degrees instead of 20 degrees. In what ways, if any, would this affect the results of the experiment? What would happen if you used 200 mL of water and heated it only 10 degreees? Can you think of any advantages or disadvantages in using either 50 mL or 200 mL of water in this experiment?

6. There are many possible sources of error in this experiment. List three that you can think of. Would each error have a <u>large effect</u>, a <u>medium effect</u>, or a <u>small effect</u> on the calculated heat content of a fuel? Also indicate whether the calculated result would be <u>too high</u>, <u>too low</u>, or could go <u>either way</u>.

Name _____   Date _____

Partner _____   Lab Section _____

Data Sheet—Experiment 11

Trial number	1	2	3	4
Fuel used				
Mass of can + water (grams)				
Mass of empty can (grams)				
Mass of water (grams)				
Final temp. of water ($^\circ$C)				
Initial temp. of water ($^\circ$C)				
Temperature change ($^\circ$C)				
Initial mass of burner (grams)				
Final mass of burner (grams)				
Mass of fuel burned (grams)				
Total heat absorbed by water (calories)				
Heat per gram of fuel (calories/gram)				

Building a Conductivity Detector and Testing for Ions

Introduction

The modern chemical laboratory has an array of devices to assist chemists in gathering information needed to solve specific problems. The equipment used by early scientists was undoubtedly hand-built by the individual scientists. In this exercise, you will build a simple electrical conductivity detector and then use it to test for the presence of ions in various liquids. Ions are an extremely important type of chemical substance and — as you will discover — they are present in nearly every common liquid that contains water.

Background Information

Many chemical substances contain separate electrically charged particles called **ions.** (See Sections 5.10 and 5.11 in *Chemistry in Context,* fourth edition, for more detail.) If ions are present, there must be equal amounts of both positively charged and negatively charged ions. As one example, when solid sodium chloride, NaCl, is dissolved in water, it breaks apart into Na^+ and Cl^- ions. Because these ions are free to move independently and because they carry electric charges, they can "conduct" electricity through the solution. This property provides a very simple and useful way to test for the presence of ions: If ions are present, the solution will conduct electricity. Furthermore, the magnitude of the conductance is directly proportional to the concentration of ionic substances dissolved in the liquid.

The device you will construct in this exercise is designed to indicate whether a solution can conduct electricity. When the probes are immersed in a conducting liquid, the light-emitting diode (LED) will light up. (Light-emitting diodes are solid state devices that emit light when electric current flows through them.) The detector is designed so that a conducting material completes the circuit and current flow causes the LED to emit visible light. The tester also works well for measuring the conductivity of solids: If the probes are touched to a solid and the LED lights up, then the solid conducts electricity.

Reference: This design for a conductivity detector was originated by F. J. Gadeck, *J. Chem. Educ.*, **64**, 628 (1987).

Overview of the Experiment

1. Gather together all the parts for the detector.
2. Practice soldering if necessary.
3. Assemble and solder together the detector.
4. Test the detector to be sure it works.
5. Use the detector to test various liquids, solutions, and materials in the laboratory.
6. Take the detector with you to test materials outside of the laboratory.

Materials Needed

Chemicals	Parts list for conductivity detector
• distilled or deionized water • 1% NaCl solution • 1% NaOH solution • 1% HCl solution • 1% sugar solution • 10% ethanol in water	• 9-V battery • battery snap-on connector • 2 12-inch lengths of wire (black and red) • 1 light-emitting diode • 2 alligator clips (optional) • 1 1,000-ohm 1/4 watt resistor • 12 inches of black electrical tape • 1 plastic film canister Tools to be shared: • scissors • wire stripper • pliers • solder gun and solder (or tape solder) • matches and/or candle (for tape solder)

The electrical circuit diagram for the conductivity detector is shown in the following diagram.

Figure 12.1 Electrical circuit diagram for the conductivity detector

Procedure for Soldering Wires Together

Solder is a low-melting mixture of several metals that is melted onto wires, which are then joined together as the solder solidifies, thus making an electrical connection. There are two options for soldering the wires in this exercise. Your instructor will indicate which one to use: (a) an electrically heated soldering iron or "gun" or (b) small strips of tape solder that are wrapped around the wires to be joined and then heated with a match or candle flame. The tape solder melts at a low temperature, thus making a soldering iron unnecessary.

Before assembling the detector, you should first practice soldering together some scrap pieces of wire as follows:

1. Using a wire stripper, strip off about 1 cm of the plastic coating from the ends of two wires. (There are several types of wire strippers. Your instructor will demonstrate the type that is available in your laboratory.)

2. Twist the bare ends of the wires together to make a mechanical connection.

3. Do ONE of the following

    a. Cut a piece of tape solder that is 0.5–1.0 cm long. Wrap the tape solder around the connected wires and then heat the tape solder gently with a match or candle. The tape solder should melt and flow over the wires, joining them together.

    b. Obtain a strip of solder wire. Switch on the soldering iron or gun and hold it against the end of the solder wire until the solder begins to melt (to confirm that the soldering iron is hot). Then touch the hot tip of the soldering iron to the wires to be joined and simultaneously hold the end of the solder wire to the heated spot. Melted solder should flow smoothly over the hot wire junction. Remove the soldering iron and wait for the joint to cool. Check with the instructor to be sure you have a good joint.

4. Repeat steps 1–3 until you can make a simple solder joint that seems solid.

I. Assembling the Detector

1. Gather the parts for the conductivity detector and be sure you can correctly identify each part.

resistor   LED

2. Prepare the film canister by punching four holes in the cap, arranged as shown below. Punch the holes from the inside of the cap to avoid damaging the lip of the cap. If a suitable hole puncher is available, use it. Otherwise, use pliers to hold a nail in a flame until it is hot (not red hot). Then use the nail to melt suitable holes in the top of the plastic film container.

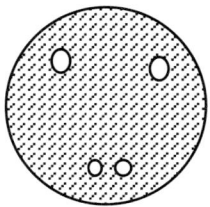

3. Push the two wires from the LED through the two closely spaced holes in the cap of the film canister.

4. Locate the longer wire of the LED. If thin plastic tubing is available, cut a length of it so that it covers all of the wire except for about 1 cm. This tubing (sometimes called "spaghetti") insulates the bare wires.

5. Twist the **long** wire of the LED and a wire from the resistor together. (**Note:** If the wrong wire is attached to the resistor, the LED will be permanently damaged when current flows through the circuit.)

6. Solder the resistor and the LED together at this joint.

7. If it is available, cut a length of thin tubing so that it covers all of the other resistor wire except for about 1 cm.

8. Solder the remaining resistor wire to the red wire from the battery clip.

9. Thread a wire through one of the remaining holes in the cap. Locate the other LED wire. If thin tubing is available, cut a length of it so that it covers all of the LED wire except for about 1 cm.

10. Solder the shorter wire of the LED to this wire.

11. Thread a second wire through the cap. Locate the black wire from the battery clip and solder it to the wire from this wire.

12. *Optional (recommended):* Attach alligator clips to the ends of the two wires extending through the cap. First make a mechanical connection for each; then solder the connections.

13. Connect the battery clip to a 9-volt battery and test the detector by touching the wires or the alligator clips together. The LED should light up.

14. If the detector does not work, check the entire system against the diagram in *Figure 12.1* to see if you have made a soldering error. Also check for loose connections. If you cannot find an error, check with your instructor.

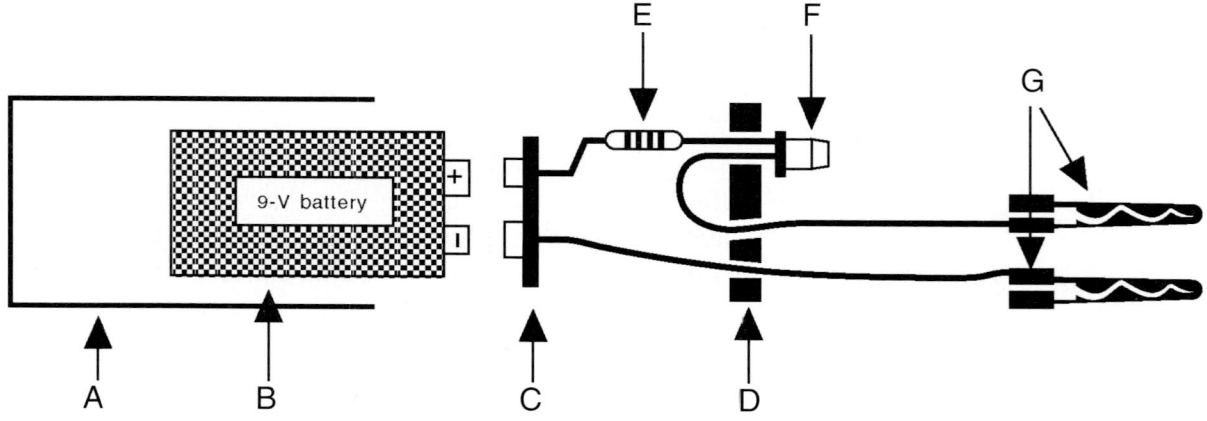

Figure 12.2 Diagram of the finished conductivity detector: (A) film canister, (B) 9-V battery, (C) battery clip, (D) canister cap, (E) resistor, (F) LED, and (G) alligator clips

15. If the detector works, package it in a black plastic film canister (*Figure 12.2*). Be careful not to pull apart any connections or have any bare wire connection touching any other bare connection. (Optional: After assembly, secure the top with black electrical tape.)

II. Using the Conductivity Detector to Test for Ions

If alligator clips are used, do not insert the metallic clips directly into the solutions. Instead, clip a short piece of wire (e.g., a partially unfolded paper clip) into each alligator clip. These wires can then be inserted into the solutions to be tested. Using wires in this fashion keeps the alligator clips out of the solutions and prevents the clips from becoming corroded.

To test a liquid sample, dip the wires attached to the alligator clips (or the wires themselves if no clips are used) into the liquid to be tested. The wires should dip in about a half an inch and should not touch each other. If the light goes on, it means that an electric current is flowing through the solution and that the solution contains ions. *It is important to rinse off the wires with pure water each time they are to be inserted into a new solution so as not to carry over contamination from one solution to the next.*

The tests that follow can be performed in small beakers, small test tubes, or in a plastic wellplate. Be sure to record your observations on the data sheet as you proceed. For each test, record whether you observe no light, a bright light, or a dim light.

1. Test the conductivity of (a) pure water (distilled or deionized) and (b) tap water. What can you conclude about the presence of ions in each?

2. Next, test a sample of salt water (1% solution of sodium chloride). Is the result what you expected?

3. Test a 1% solution of sodium hydroxide (formula: NaOH). Does it contain ions?

4. Hydrochloric acid is a water solution of HCl (which, by itself, consists of HCl molecules). Test a 1% solution of hydrochloric acid. What do you conclude? How would a chemist describe what is present in hydrochloric acid solutions?

5. Test a 1% solution of sugar in water. Does it contain ions?

6. Test a 10% solution of ethanol (ethyl alcohol) in water. Does it contain ions?

7. Using salt water, find out whether your conductivity tester is sensitive to the length of the wire that is immersed in the liquid being tested. Also find out whether the sensitivity of your detector is affected by the distance between the wires that are immersed in a liquid.

8. Devise and carry out an investigation to determine how sensitive your detector is to the presence of ions in solution. Record your method and the results.

9. **Optional:** Your instructor may provide additional solutions to test in the laboratory. If so, be sure to write down what the solutions are and what you observe.

10. Now that you know how to use the conductivity tester, take it with you when you leave the lab and test at least five liquids that you encounter during the course of a day. Enter the data in the data sheet. Be prepared to report your observations at the next meeting of your class or lab section. Some suggestions for liquids to try include (a) tap water, (b) a stream or lake, (c) sweat, (d) urine, (e) foods such as coffee, tea, fruit juice, milk, soft drinks, a potato, a piece of apple or other fruit.

> **CAUTION:** For testing beverages and foods, always put a small amount of the substance in a separate clean container. *Never dip the tester into any food or beverage that you will later ingest.*

Questions To Be Answered After Completing This Experiment
Write out answers to the following questions on a separate sheet of paper and hand it in along with the data sheet.

1. Which of the tested liquids contained ions? Which ones did not contain ions?

2. Did the brightness of the light vary from solution to solution? Cite specific examples. What does a dim light tell us about the number of ions in a solution?

3. If you wanted to make a solution of an unknown solid substance to test it for the presence of ions, would it matter whether you used pure water or tap water to make the solution? Explain briefly.

4. If you were able to test some commercial products, read the labels on the containers of these liquids and attempt to determine what substances were responsible for the liquid's conductivity, if any.

5. The purity of ultrapure water is often measured with conductivity detectors. Based on what you have observed, why is this a good test for water purity? In what way might it be incomplete?

6. Describe the results you got when you varied the length of wire immersed and the distance between wires. Offer explanations for the behavior you observed.

Name _____   Date _____

Partner _____   Lab Section _____

Data Sheet—Experiment 12

Tests of Liquids in the Lab

Liquid	Conductivity
Pure water (distilled or deionized)	
Tap water	
Salt water	
Sodium hydroxide solution	
Hydrochloric acid	
Sugar solution	
Ethanol solution	

Common Liquids Tested Outside the Lab

Liquid	Conductivity

Effect of varying the length of wire immersed.

Effect of varying the distance between the wires.

Name _____   Date _____

Partner _____   Lab Section _____

Data Sheet—Experiment 12, page 2

Write out the procedure you used to test the sensitivity of your detector. Include your results and conclusions.

Results of optional tests.

Analysis of Vinegar

Introduction

This laboratory assignment is an introduction to the titration method of chemical analysis. You will use the titration method to determine the acid content in a familiar substance, vinegar. Since later laboratory assignments will use the titration method for other chemical analyses, the background discussion and the discussion of the procedure are quite detailed.

Background Information

Acids and bases are among the most common of chemical substances. Chapter 6 in *Chemistry in Context* has an extensive discussion of acids and bases. Solutions of **acids** in water contain excess H^+ ions, while solutions of **bases** contain excess OH^- ions (see text, Sections 6.1 and 6.2). Vinegar is a dilute solution of acetic acid, $HC_2H_3O_2$, in water.

A characteristic property of acids and bases is that they react with each other. Thus, for example, acetic acid in vinegar will react with sodium hydroxide (a base) to produce water and sodium acetate:

$$\text{acetic acid} + \text{sodium hydroxide} \rightarrow \text{water} + \text{sodium acetate}$$
$$HC_2H_3O_2 + NaOH \rightarrow H_2O + NaC_2H_3O_2$$
$$\text{(an acid)} \quad\quad \text{(a base)}$$

This type of reaction is the basis for the **titration method of analysis** for acids. In a titration, a known quantity of acid solution is measured out; then a solution of base (usually sodium hydroxide, NaOH) is added slowly until just enough has been added to react with all of the acid. If one knows the volume of NaOH added and its concentration, it is possible to calculate how much acid must have been present to react with the NaOH. Finally, if the volume of the acid solution is known, it is possible to calculate the concentration of the acid. Concentrations are expressed as **molarity** (M), which is defined (Section 6.4 in the text) as the number of moles of a substance in exactly 1 liter of solution.

In a titration, an **indicator** is added. This is a chemical substance that has one color in the presence of excess H^+ ions and a different color when there are excess OH^- ions. The color change signals the end of the titration, i.e., the point where just enough NaOH has been added to completely react with all of the acid. The indicator used in this experiment is *phenolphthalein*, which is colorless in acid and red in base. To do a titration, a drop of the indicator is added to a measured amount of acid. Then NaOH solution is added, one drop at a time, until one drop of NaOH changes the solution from colorless to pink. This is the "end point" of the titration. Indicators are intensely colored so only one drop is needed for a titration.

For accurate calculations, the exact volumes of the acid and base must be known. However, if we assume that the plastic pipets used in the experiment deliver drops of a uniform size, then instead of measuring exact volumes, you can count drops of each solution to obtain the volumes of acid and of base used. If the number of drops for each reactant is known and the concentration of sodium hydroxide is known, it is possible to calculate the concentration of acid in the vinegar solution.

Overview of the Experiment

1. Obtain a 24-well wellplate, plastic transfer pipets, and the necessary solutions.
2. Practice dispensing drops evenly.
3. Practice titrating vinegar with sodium hydroxide solution.
4. Titrate samples of vinegar to determine their acetic acid content.
5. Calculate the molarity and percent acidity of the acetic acid in the vinegar.

Materials Needed

Chemicals	Equipment
• vinegar samples • sodium hydroxide solution of known concentration • indicator solution (phenolphthalein)	• 24-well wellplate • 4 plastic transfer pipets • small stirring rod or toothpick • piece of white paper • tape or labels for the pipets

EXPERIMENTAL PROCEDURE

SAFETY NOTE: Sodium hydroxide solution is corrosive to your skin and can cause serious eye damage. Therefore, you MUST wear safety glasses or goggles at all times and should try to avoid getting any of the solution on your hands. (If you do get it on your hands, wash the skin immediately with plenty of water.)

I. Getting Organized

1. Obtain a plastic wellplate with 24 wells and a small stirring rod or toothpick. Place the wellplate on a sheet of white paper so that color changes can be seen easily. Your wellplate probably has letters and numbers to identify the wells. If not, refer to the diagram in *Figure 1.2*. These labels are used in the following instructions.

2. Label three plastic pipets as follows: vinegar, NaOH, indicator.

3. Fill the "vinegar" pipet from a bottle of vinegar and record on the data sheet the brand of vinegar.

4. Fill the "NaOH" pipet from the supply bottle. Record on the data sheet the exact concentration of the NaOH solution (which should be written on the label of the supply bottle).

5. Partially fill the "indicator" pipet with phenolphthalein solution.

II. Pipet Practice

Fill the unlabeled plastic pipet with water and try dispensing drops into a well of the wellplate. Your goal is to be able to confidently add a known number of drops to the bottom of the well, one drop at a time. Squeeze the bulb gently, while the pipet is held vertically and directly over the center of the well. You may find it helpful to use two hands to steady the pipet. Plastic transfer pipets easily acquire air bubbles in their stems that lead to frustrating "half-drops" and "quarter-drops" that introduce errors. The best solution is to reserve one well (e.g., D6) for solution waste. During a titration, "half-drops" are not added to the titration well or counted. They are simply added to the waste well. When you can confidently add a known number of drops to a well, proceed to the practice titration.

III. Practice Titration

1. Add 10 drops of vinegar to well A1 in the wellplate.

2. Then add 1 drop of indicator. The mixture should be colorless.

3. Start adding NaOH, <u>one drop at a time</u>, (counting the drops), with gentle stirring. Carefully observe what happens. As each drop is added, a slight pink color may be visible, which should disappear upon stirring. After enough drops have been added, the solution in the well will turn pink and stay pink. Make a note of how many drops were used. Then add a few more drops of NaOH, one drop at a time, with stirring. The color will likely become darker red as the solution becomes more and more basic.

IV. Measurement of the Acetic Acid Content of Vinegar

In doing titrations, your goal is to catch the point where <u>one drop</u> of NaOH produces the first permanent color change. This is the "end point" where all of the acid has been used up by reaction with the NaOH.

1. Carefully add <u>20 drops</u> of vinegar (count them) to each of six wells (B2, B3, etc.). Make sure that the drops fall directly to the bottom of the well and are not "trapped" along the side. If you make a mistake, simply start again in another well. Record the position numbers for the wells you are using (B2, B3, etc.) on the data sheet.

2. Add <u>1 drop </u>of indicator to each well.

3. Start adding NaOH, <u>one drop at a time</u>, to the first well as you stir the solution. As you get close to the stopping point, you will find that the momentary pink color persists longer after each drop is added. Stop when one drop of NaOH produces a permanent pink or red color. Record the number of drops you used on the data sheet. If you make a mistake, skip this trial and go on to the next one.

4. Repeat this titration for each of the other five wells. With a little practice, the titrations go very fast. Be sure to record the number of drops used for each well on the data sheet.

> **Hint:** Once you know approximately how many drops of NaOH are needed, you can quickly add about 2/3 of that amount. For example, if the first titration required 15 drops of NaOH, then for the next trial, you could quickly add about 10 drops of NaOH and then add the remaining drops slowly.

> **Note:** If you miss the end point, or lose count of drops or are uncertain of the result for any other reason, it is good scientific practice **not** to erase the data. Simply make a note in the table as to what you think went wrong, draw a diagonal line through that column, and do another titration. All scientists make mistakes when doing lab work! Results crossed out will not affect your grade.

5. You should now have six results. They ought to agree with each other within a few drops. If they do not, consult your instructor.

6. **Optional extension:** Analyze another kind of vinegar. First, rinse the vinegar pipet with the new sample by twice filling the pipet and emptying it into a sink or waste container. Use a new set of wells in the wellplate for the new sample.

Clean Up

The contents of the wellplate can be safely emptied into a sink. Rinse the wellplate with tap water, then with distilled or deionized water if available. Leave it upside down on a paper towel to drain. Your instructor will tell you what to do with the plastic pipets.

V. Calculations

NOTE: A pocket calculator will be quite helpful for these calculations.

The logic behind the calculations for a titration analysis is a little tricky. Even if you do not understand the explanation completely, the actual calculations are quite easy.

The chemical equation for the reaction of acetic acid with sodium hydroxide is

$$HC_2H_3O_2 + NaOH \rightarrow NaC_2H_3O_2 + H_2O.$$

This equation shows that *1 mole of acetic acid reacts with 1 mole of sodium hydroxide*. Thus, for the titration of vinegar, the number of moles of NaOH added from the pipet is exactly equal to the number of moles of acetic acid (from vinegar) in the well. Expressed mathematically, we can say that

moles of NaOH added from the pipet = moles of $HC_2H_3O_2$ in the well   **(1)**

The number of moles of NaOH in any sample of NaOH solution can be calculated from the molarity of the NaOH (expressed as moles/liter) and the volume (expressed as liters), using the following equation:

$$\text{moles of NaOH} = \frac{\text{moles of NaOH}}{1 \text{ liter of NaOH solution}} \times \text{liters of NaOH solution} \qquad (2)$$

A similar equation can be written for the moles of acetic acid in the vinegar sample.

$$\text{moles of } HC_2H_3O_2 = \frac{\text{moles of } HC_2H_3O_2}{1 \text{ liter of vinegar}} \times \text{liters of vinegar} \qquad (3)$$

By substituting into the first equation, we can write the following equation:

$$\frac{\text{moles of NaOH}}{1 \text{ liter of NaOH}} \times \text{liters of NaOH} = \frac{\text{moles of } HC_2H_3O_2}{1 \text{ liter of vinegar}} \times \text{liters of vinegar} \qquad (4)$$

Or, writing this more succinctly,

molarity of NaOH $\times$ liters of NaOH = molarity of $HC_2H_3O_2 \times$ liters of vinegar   **(5)**

This equation can be rearranged to put it into a more useful form:

$$\text{molarity of } HC_2H_3O_2 = \text{molarity of NaOH} \times \frac{\text{liters of NaOH}}{\text{liters of vinegar}} \qquad (6)$$

There is one small problem: You don't know the actual volumes in liters or milliliters; you know only the number of drops of each solution that you used. If we assume that all of the drops from both pipets have the same volume, then the ratio of volumes expressed as drops should be the same as the ratio of volumes expressed as liters. This finally gives us an equation for calculating the molarity of acetic acid:

$$\text{molarity of } HC_2H_3O_2 \;=\; \text{molarity of NaOH} \times \frac{\text{no. of drops of NaOH}}{\text{no. of drops of vinegar}} \qquad (7)$$

To do the calculations, first look carefully at the six or more titration results in the data table. Does any one result seem way out of line from the others? If so, you probably did something wrong, and it is legitimate to omit that result. Simply write "omit" underneath it. Using all of the results that you think are valid, calculate the average number of drops of NaOH and record this on the data sheet.

Next, calculate the molarity of acetic acid in the vinegar, using equation (7). Multiply the molarity of the NaOH by the average number of drops of NaOH and then divide by 20, the number of drops of vinegar used.

Labels on vinegar bottles usually specify the *percent acidity,* which is defined as the number of grams of acetic acid per 100 mL of vinegar. One mole of acetic acid weighs 60 grams; therefore, multiplying the molarity (moles/liter) by 60 will give the number of grams of acetic acid per liter and then dividing by 10 will give the number of grams per 100 mL (since 100 mL = 1/10 of a liter). Using this information, calculate the percent acidity for your vinegar sample and record it on the data sheet.

An example will make this clearer. Suppose you found that the acetic acid concentration was 0.55 mole per liter of vinegar. Then the calculations would be as follows:

$$\frac{0.55 \text{ mole } HC_2H_3O_2}{1 \text{ liter vinegar}} \times \frac{60 \text{ g. } HC_2H_3O_2}{1 \text{ mole } HC_2H_3O_2} = \frac{33 \text{ g. } HC_2H_3O_2}{1 \text{ liter vinegar}}$$

$$\frac{33 \text{ g. } HC_2H_3O_2}{1 \text{ liter vinegar}} = \frac{33 \text{ g. } HC_2H_3O_2}{1{,}000 \text{ mL vinegar}} = \frac{3.3 \text{ g. } HC_2H_3O_2}{100 \text{ mL vinegar}} = 3.3\% \text{ acetic acid}$$

Report your result in whatever manner is specified by your instructor.

Questions To Be Answered After Completing This Experiment

Write out answers to the following questions on a separate sheet of paper and hand it in along with the data sheet.

---

1. During the experiment, you were cautioned about making sure that all drops of solutions were at the bottom of the well, not along the sides. Why?

2. Suppose that during the titration, you added sufficient NaOH to change the color of the indicator to a darker pink. How would this change the results? Explain.

3. What is the advantage of doing several trials, especially since you are doing exactly the same thing each time?

4. Could you have done the titration "backwards," i.e., start with 20 drops of NaOH, add phenolphthalein indicator, and then add the vinegar one drop at a time? Explain. What would you observe?

5. Can you think of ways to improve this method of analysis? What changes might be possible to give you more confidence in the accuracy of the results?

6. Why do you suppose that the acetic acid content in vinegar is reported as percent acidity rather than as molarity?

7. Suppose you were asked to determine the exact acetic acid concentration in an acetic acid solution that was about 50% acid. How would you change the procedure you used to determine the acetic acid content of vinegar so it could be used to solve this problem?

Name _____   Date _____

Partner _____   Lab Section _____

Data Sheet—Experiment 13

Molarity of the NaOH solution: _____ mole/liter

Description of the vinegar sample: _____

Trial #	1	2	3	4	5	6	7	8
Well number								
Drops of vinegar								
Drops of NaOH								
Comments								

(Extra columns are provided in case they are needed.)

Average number of drops of NaOH: _____ (per 20 drops of vinegar)

Calculated molarity of acetic acid: _____ mole/liter   **(Eq. 7)**

Percent acidity in the vinegar: _____ %

Measurement of Water Hardness

Introduction

Water from lakes, rivers, or wells that contains significant concentrations of calcium ions (Ca^{2+}) and magnesium ions (Mg^{2+}) is known as **hard water**. These ions are responsible for an assortment of problems ranging from aesthetic ones, such as bathtub rings and soap scum, to more critical ones such as plugged steam lines and damaged hot water heaters. Calcium ions get into water if the water comes in contact with limestone (calcium carbonate); magnesium ions likewise enter water when ground water passes through minerals that contain magnesium ions.

In this experiment, you will analyze water samples for the combined concentrations of calcium and magnesium ions known as "total hardness." You will use a titration method of analysis, which is described in the text, Section 5.15.1. The titration method is fast, easy, and one of the most widely used analytical techniques in many areas of chemistry and biology.

Background Information

The analysis for water hardness is based on the chemical reaction of Ca^{2+} (and Mg^{2+}) with an ion called dihydrogenethylenediaminetetraacetate, $C_{10}H_{12}N_2O_8{}^{2-}$. Since the name and formula are complicated, it is almost always called simply "EDTA" (and often written as "H_2EDTA^{2-}"). The reaction involves one calcium (or magnesium) ion reacting with one H_2EDTA ion. (Note that in this reaction, one Ca^{2+} ion replaces two H^+ ions.)

$$Ca^{2+} + H_2EDTA^{2-} \rightarrow Ca(EDTA)^{2-} + 2\,H^+$$

The analytical procedure is a **titration.** A known amount of water is measured out, and EDTA solution is added slowly until just enough has been added to completely react with all of the calcium and magnesium ions in the sample. In order to know when enough EDTA solution has been added, a small amount of an **indicator** called calmagite is used. At pH 10 (an alkaline solution), the indicator by itself has a deep blue color, but in the presence of metal ions, it changes to red. Thus, if a drop of indicator is added to a solution containing calcium and magnesium ions, the solution will have a reddish color. When enough EDTA has been added to react with all of the calcium and magnesium, the solution will turn blue.

The strategy of the experiment will be to first use an EDTA solution to titrate a reference calcium solution of known concentration and then use the same EDTA solution to titrate an unknown water sample. By comparison of the two sets of results, it is easy to calculate the hardness of the unknown sample.

For simplicity, the amounts of Ca^{2+} and Mg^{2+} that are present in the water sample will be lumped together. Furthermore, the common practice in water hardness is to report the results as *milligrams of CaCO3 per liter of water*. This does not mean that solid calcium carbonate is actually present, but rather that this amount of solid calcium carbonate could be formed from the amount of Ca^{2+} (and Mg^{2+}) present in the water.

Overview of the Experiment

1. Test the indicator color change.
2. Do 3 or 4 titrations of water containing a <u>known</u> concentration of calcium ions.
3. Calculate the calibration factor for the EDTA solution.
4. Do 3 titrations for each of several water samples containing an <u>unknown</u> concentration of calcium and magnesium ions.
5. Calculate the water hardness for the unknown samples.

Materials Needed

Chemicals	Equipment
• EDTA solution	• 24-well wellplate
• pH 10 buffer solution	• 2 or 3 graduated-stem plastic pipets
• calmagite indicator solution	• 2 additional plastic pipets of any style
• distilled water	• Small stirring rod or toothpick
• reference solution of known hardness (500 mg $CaCO_3$ per liter)	• 2 small test tubes + a rack to hold them (alternate: 2 small beakers)
• water samples of unknown hardness	• sheet of white paper

EXPERIMENTAL PROCEDURE

I. Preliminary Work: Getting Organized

1. Obtain 2 small test tubes that are <u>clean</u> and <u>dry</u> and a rack to hold them. Fill them about 1/3 with the following solutions. Be sure to label them so that you know which one is which.
    a. EDTA solution
    b. Reference solution of known hardness (equivalent to 0.500 mg of CaCO3 per mL of water)

2. Obtain 4 plastic graduated-stem pipets that are clean and dry.
    a. Label one as "buffer" and fill it 1/3 full with "buffer solution, pH 10."
    b. Label another as "indicator" and fill it 1/3 full with "calmagite indicator."
    c. Label a third pipet as "EDTA" and place it in the EDTA test tube.
    d. Label a fourth pipet as "Ca reference" and place it in the reference solution.
    Be sure that all pipets are labeled so that you will not get them mixed up.

3. Obtain a 24-well plastic wellplate and a short stirring rod made of glass or plastic. Place the wellplate on a piece of white paper (to aid in seeing the color changes). Note the letters and numbers on the wellplate (or refer to the diagram in *Figure 1.2*). Have the data page of this experiment ready to record numbers as you obtain them.

4. <u>Pipet practice</u>. (If you have done a previous titration experiment with this kind of equipment, e.g., Experiment 13, *Analysis of Vinegar,* you can skip this step.)

   Take an empty graduated-stem plastic pipet, fill it with water, and try dispensing drops into a well of the wellplate. Your goal is to be able to confidently add a known number of drops to the bottom of the well, one drop at a time. Squeeze the bulb gently while the pipet is held vertically and directly over the center of a well. You may find it helpful to use two hands to steady the pipet. These pipets easily acquire air bubbles in their stems, which lead to frustrating "half-drops" and "quarter-drops" that introduce errors. The best solution is to reserve one well for solution waste. During a titration, "half-drops" are not added to the titration well or counted. They are simply added to the waste well. When you can confidently add a known number of drops to a well, proceed to the practice titration.

II. Titrations of the Reference Solution

1. <u>Test the color change of the indicator.</u> To well A1, add 8–10 drops of pure water. Then add 2 drops of pH 10 buffer and 1 drop of calmagite indicator. Stir with a clean stirring rod. The solution should be a bright, clear blue. If <u>not</u>, add 1 drop of the EDTA solution to produce a blue color. This is the color you will be looking for at the end of each titration. It should be *pure blue* with no trace of red. (If not, check with your instructor before proceeding further.)

2. Rinse and fill a 1-mL graduated-stem plastic pipet with the calcium reference solution. Practice until you can fill it *exactly* to the 1-mL line. (The best method for doing this is to gradually squeeze air out of the bulb, release the pressure, and observe how high the solution rises. Repeat until the solution just rises to the 1-mL mark.) Dispense exactly 1 mL of this solution into each of four wells in the wellplate (e.g., wells A2–A5).

3. To the first well, add 2 drops of buffer <u>plus</u> 1 drop of indicator (in that order). Then start adding EDTA solution slowly, <u>one drop at a time</u>, counting drops and stirring after each addition. Continue until the color <u>starts</u> to change from red to purple. Wait a few moments to see if the color continues to change (the reaction is slow); then *slowly* add additional drops, stirring after each one, until the color becomes pure blue. <u>Keep track of the total number of drops added</u>. When you reach the end point, enter in the table the number of drops used.

4. Repeat the same procedure with the samples in the other three wells, titrating one sample at a time. Remember to add 2 drops of buffer and 1 drop of indicator (always in that order) before each titration.

5. Do additional titrations if you are uncertain of any of the four titrations (for example, if you lost any drops of solution or think that you went past the end point or lost count). With a little practice, the titrations can be done quite rapidly. (The slowest part of this particular titration may be waiting for the chemicals to react completely.)

III. Titrations of Unknown Water Samples

Obtain one or more water samples to be analyzed, as directed by your instructor. These might be the tap water in your laboratory or in your dormitory or home. They might be samples from a nearby lake or river. They might be from a well or a spring.

> **Note:** Some water samples – especially tap water in some buildings – may not give a good color change to pure blue. This is caused by traces of iron or copper in the water. If you find you have such a sample, check with the instructor before proceeding further. It may be necessary to titrate to a purple color rather than a pure blue.

1. Obtain a clean 1-mL graduated-stem pipet. Label it for water samples.

2. Carefully dispense 1 mL of a water sample into a clean well in the wellplate—e.g., B1.

3. Add 2 drops of buffer followed by 1 drop of calmagite indicator. Proceed to titrate with the EDTA solution in exactly the same way you did in Part I. Add the EDTA one drop at a time, keeping track of the number of drops added. Stir after each addition, watch for the start of the color change and then continue slowly until one drop turns the solution pure blue. Because this is an unknown solution, you have no idea the first time whether it will take more or fewer drops of EDTA than the reference solution did; thus, you will need to do the first titration slowly. Record the total numbers of drops of each solution used in the data table.

4. Do two or three additional titrations for this sample. Depending on how much EDTA was used in the first trial, you may want to use either a larger or smaller volume of the unknown sample for the remaining titrations. (If you do use a different volume of water, be sure to record this in the data table.) Remember to add the buffer and indicator each time. Record the numbers in the data table.

5. **Optional**: If you have time, you may be curious to analyze additional water samples or your instructor may assign samples to be analyzed. Remember to rinse the sample pipet thoroughly with each new sample before measuring out 1-mL portions of that sample. Better yet, use a new, clean pipet.

6. **Optional: Softening of Water by Ion Exchange.** Take about 10–20 mL of water that is relatively "hard" and allow it to run slowly through a small tube containing beads of an ion exchange resin (Section 5.15.1 in the text). Collect some of the softened water in a clean test tube and titrate 1-mL portions of the water using the same procedure as above. Is the hardness of the water different? Give a simple chemical explanation of what happened in the tube.

7. **Optional: Softening of Water by Distillation**. A simple distillation apparatus may be set up in the laboratory. See Figure 5.29 in *Chemistry in Context*, fourth edition, for a similar apparatus. Tap water or other hard water will be placed in the heated flask. Obtain a few milliliters of the distilled water in a clean test tube and titrate 1-mL portions of this water using the same procedure as above. How does the hardness of distilled water compare with the starting material?

8. **Optional: Measurement of Calcium and Magnesium by Instrumental Methods**. Your instructor may ask you to analyze some water samples for calcium and magnesium using modern analytical instrumentation. The instrument is first calibrated using solutions of known concentration and then unknown solutions are analyzed. You will be given detailed instructions

in how to use the instrument and the calculations. You can then compare the instrumental results with those obtained by the titration procedure described previously.

IV. Calculations

Note: The calculation procedure is slightly different from the one described in the text (Section 5.15.1). For the procedure used here, you do not need to know either the molarity of the EDTA or the volume in milliliters of EDTA used per titration.

1. Look carefully at the data from each of the sets of titrations to see whether the results show consistency or whether any one result in a given set should be eliminated because it appears to be an "outlier." If so, make a note beside it in your notebook.

2. Calculate the average number of drops of EDTA used for each set of titrations, omitting any results that you think are not valid.

3. In order to do the calculations, you need to know the "hardness" of the reference solution. This should be written on the bottle or posted in the lab. Ideally it will be 500 mg/liter, which is the same as 0.500 mg/mL. (Remember that hardness is normally expressed as mg of $CaCO_3$ per liter.)

4. For the first set of titrations (with the reference solution), divide the milligrams of hardness per milliliter by the average number of drops of EDTA used.

$$\frac{\text{mg hardness in ref. sol'n}}{1 \text{ mL ref. sol'n}} \times \frac{1 \text{ mL of ref. sol'n}}{\text{ave. drops of EDTA}} = \frac{\text{mg of hardness}}{1 \text{ drop of EDTA}} \qquad (1)$$

This gives the mg of hardness corresponding to 1 drop of EDTA solution. This is the <u>calibration factor</u>. It is unique for the particular EDTA solution and your equipment.

5. For the analysis of an unknown water sample, multiply the average number of drops of EDTA used by the calibration factor to obtain the mg of hardness per mL of water.

$$\frac{\text{drops of EDTA}}{1 \text{ mL sample}} \times \frac{\text{mg of hardness}}{1 \text{ drop of EDTA}} = \frac{\text{mg of hardness}}{1 \text{ mL sample}} \qquad (2)$$

6. Finally, convert this into milligrams of hardness *per liter* of water. This is the normal way of reporting water hardness.

$$\frac{\text{mg of hardess}}{1 \text{ L sample}} = \frac{\text{mg of hardness}}{1 \text{ mL sample}} \times \frac{1000 \text{ mL}}{1 \text{ L}} \qquad (3)$$

7. Rate the hardness of each water sample analyzed, using the following classification.

Description	Concentration
Very hard	Over 300 mg/L of calcium carbonate
Hard	150 to 300 mg/L of calcium carbonate
Moderately hard	50 to 150 mg/L of calcium carbonate
Soft	0 to 50 mg/L of calcium carbonate

V. Sharing Your Results with Other Students

Your instructor will provide details on how to share your results with other students. If everyone has analyzed the same sample (e.g., tap water), then results can be assembled and compared. If the class was divided up to analyze various samples, you should record what other students found for the other samples.

Questions To Be Answered After Completing This Experiment

Write out answers to the following questions on a separate sheet of paper and hand it in along with the data sheet.

---

1. Based on your own data and/or class data, how would you classify the hardness of your local tap water?

2. If your class analyzed other water samples (lake, river, well water, etc.), did you find significant differences in hardness? If so, suggest possible explanations.

3. If you investigated the use of an ion exchange resin, how effectively did it soften the water? What were the levels of hardness before and after treatment?

4. If you analyzed a sample of water from a distillation apparatus in the lab, what did you conclude about its effectiveness in removing calcium and magnesium ions?

5. In these titrations, there may be an uncertainty of at least one drop in identifying the exact end point. If you use 20 drops of EDTA solution in a titration, what percent uncertainty in the hardness is contributed by adding one extra drop?

$$\% \text{ error} = \frac{\text{no. of drops you are in error}}{\text{no. of drops used}} \times 100\%$$

6. What household problems are caused by hard water? See textbook, Section 5.15.1.

7. Is very soft water the same as very pure water? Explain.

Name _____   Date _____

Partner _____   Lab Section _____

Data Sheet—Experiment 14

Part II: Titrations of the Reference Solution

Hardness of the reference solution: _____ mg $CaCO_3$ per 1 mL

Trial	#1	#2	#3	#4	#5
mL of reference sol'n	1.00	1.00	1.00	1.00	
Drops of EDTA					

Average drops of EDTA sol'n = _____ per 1 mL of reference solution

Calibration factor = _____ mg of hardness per 1 drop of EDTA   **(Eq. 1)**

Part III: Titration of Water Samples

Description of Sample #1: _____

Titration	#1	#2	#3	#4	#5
mL of water sample	1.00				
Drops of EDTA					

(Extra columns are provided in case they are needed)

Average drops of EDTA sol'n: _____ per 1 mL of water

mg of hardness per 1 mL of water:_____   **(Eq. 2)**

mg of hardness per liter of water: _____   **(Eq. 3)**

Rating of water hardness: _____

(Space is provided on the reverse side of this sheet for up to 2 more water samples or studies of water softening. For any additional analyses, use this same format and record the information on an additional sheet of paper, which should be attached to this data sheet when submitted to the instructor.)

Name _____   Date _____

Partner _____   Lab Section _____

Data Sheet—Experiment 14, page 2

Description of Sample #2: _____

Titration	#1	#2	#3	#4	#5
mL of water sample	1.00				
Drops of EDTA					

(Extra columns are provided in case they are needed)

Average drops of EDTA sol'n: _____ per 1 mL of water

mg of hardness per 1 mL of water:_____   **(Eq. 2)**

mg of hardness per liter of water: _____   **(Eq. 3)**

Rating of water hardness: _____

Description of Sample #3: _____

Titration	#1	#2	#3	#4	#5
mL of water sample	1.00				
Drops of EDTA					

Average drops of EDTA sol'n: _____ per 1 mL of water

mg of hardness per 1 mL of water:_____   **(Eq. 2)**

mg of hardness per liter of water: _____   **(Eq. 3)**

Rating of water hardness: _____

Measurement of Chloride in River Water

Introduction

Chloride is a negatively charged ion found in water and sewage. The chloride content of natural (unpolluted) surface waters depends on the geology of the area. In areas where surface water normally has very little chloride, a higher chloride concentration implies a source from human activity. In this experiment, you will measure the chloride concentration in water samples taken from streams or rivers in your local area and also, if possible, samples from your local wastewater treatment plant. By dividing up the class to analyze different samples, it will be possible to collect a large quantity of data in a short time. You will then use the results to draw conclusions about the human impact on local surface waters.

Note that chloride, Cl⁻, (an ion) is quite different from chlorine, Cl_2, (a gaseous molecule). Chlorine is added as a disinfectant to drinking water in small amounts (less than 1 ppm) and to swimming pools in somewhat larger amounts (about 5 ppm). Chlorine will produce some chloride as it reacts, but in most cases, it is not a major source of chloride.

Background Information

Solutions containing chloride ions (Cl^-) will react with silver nitrate ($AgNO_3$) to form an insoluble white compound, silver chloride (AgCl). (Other ions that are present in the water do not participate in this reaction.)

$$AgNO_3(aq) + Cl^- \text{ (in water)} \rightarrow AgCl(s) + NO_3^-(aq) \qquad \textbf{(1)}$$
$$\text{(white)}$$

This reaction is the basis for the titration method of analysis for chloride ions. In a **titration**, a known volume of a water sample containing chloride ions is measured out, and then a solution of $AgNO_3$ is added slowly until just enough has been added to react with all of the chloride in the sample. If the volume of $AgNO_3$ added and its concentration are known, it is possible to calculate how much chloride must have been present to react with all the $AgNO_3$. Finally, if the volume of the water sample is known, the concentration of chloride ion in the sample can be calculated. Concentrations are expressed as **molarity** (M), defined as *moles of substance per liter of solution*, but in this analysis (as in many other analyses), the final results will be expressed in different units.

In order to know when enough silver nitrate has been added, a small amount of an **indicator** is added. A solution of sodium chromate (Na_2CrO_4) is used, in which chromate ions (CrO_4^{2-}) serve as the indicator. Chromate ions are yellow, but they react with silver ions to form a red precipitate of silver chromate (*Equation 2*, next page). As silver nitrate is added, the chloride is precipitated as white silver chloride (*Equation 1*). After all of the chloride has all been removed, silver ions (Ag^+)

will react with CrO_4^{2-} to form a red insoluble precipitate of silver chromate, Ag_2CrO_4 (*Equation 2*). The appearance of this red precipitate signals the end of the titration.

$$2\,Ag^+(aq)\ +\ CrO_4^{2-}(aq)\ \rightarrow\ Ag_2CrO_4(s)\ \ \text{(red precipitate)} \tag{2}$$
$$\text{(red)}$$

To do a titration, *1 drop* of indicator is added to a measured volume of water, then $AgNO_3$ is added, one drop at a time, until one drop of silver nitrate changes the solution from yellow to orange. (Orange is a combination of yellow and red.)

For greatest accuracy, the exact volume of each solution used must be known. However, if we assume that the plastic transfer pipets deliver drops of a uniform size, then instead of measuring exact volumes, you can count drops of each solution to obtain the volume of water sample and the volume of silver nitrate solution. If the number of drops of each reactant is known and the molarity of $AgNO_3$ is known, it is possible to calculate the concentration of chloride ions in the water sample.

Overview of the Experiment

1. Collect water samples at various locations.
2. Practice doing chloride titrations with tap water and silver nitrate.
3. Calculate the concentration of chloride in the tap water sample.
4. Titrate the collected water samples with silver nitrate.
5. Calculate the concentration of chloride in the collected samples.
6. Compare class data for different water samples.
7. Draw conclusions about human impact on the local aquatic environment.

Materials Needed

Chemicals	Equipment
• silver nitrate solution of known concentration • sodium chromate indicator • water samples for analysis	• 24-well wellplate • 2 graduated-stem plastic pipets (or other style plastic pipets) • small stirring rod or toothpick • 2 small beakers or other containers for $AgNO_3$ and water • piece of white paper

SAFETY NOTES

• Silver nitrate will stain your skin black, so be careful not to get it on your hands. The stain is harmless and will wear off in a few days, but in the meantime, its appearance is unattractive.

• Sodium chromate is considered to be a carcinogen.

EXPERIMENTAL PROCEDURE

I. Collect Water Samples

An important aspect of the experiment is sample collection and choice of sampling locations. Before the lab class, your instructor will assign a sample location for each pair of students (or will arrange to have samples collected and made available in the laboratory). If a major river or stream flows through your town or city, then water samples should be collected from the river upstream of the local city, from the river in the middle of the city, and downstream of the city far enough away that any sewage treatment plant effluent has freely mixed into the river. Samples should also be collected from the sewage treatment plant itself: both the effluent and the influent (which should be aerated because of its high suspended solid content). Your class also needs to have samples from the municipal drinking water sources, such as reservoirs or lakes. Finally, it is helpful, for comparison purposes, to collect samples from sources that are "clean," such as ponds or small streams, that do not have any likely human impact. If you collected your own sample(s), be sure to record accurately the location from which you sampled the water and label the sample(s).

II. Get Organized for Titrations

1. Obtain a plastic wellplate with 24 wells and a small stirring rod. Place the wellplate on a sheet of white paper so that color changes can be seen easily.

2. Obtain two graduated-stem plastic pipets (or other matched-style pipets). Label one "water sample" and the other "silver nitrate" (or "AgNO3"). *Labeling is crucial because it is easy to get them mixed up!*

3. Using a clean *dry* beaker or other container, obtain about 10 mL of silver nitrate solution. Label the container and record the exact concentration (given on the label of the supply bottle) on the data sheet.

4. Obtain one or more water samples for analysis. These may be samples you have collected yourself (as described above) or samples provided by the instructor. You will start by analyzing tap water.

III. Practice the Titration

1. Obtain some tap water in another beaker. Using the "water sample" pipet, add 10 drops of this tap water to well A1 in the wellplate.

2. Add 1 drop of sodium chromate indicator. The mixture should be pale yellow-green.

3. Start adding $AgNO_3$ solution, <u>one drop at a time</u> (counting the drops), with gentle stirring. Carefully observe what happens. As each drop is added, you will observe a cloudy appearance as solid silver chloride is formed; then a reddish color may appear that will disappear on stirring. After a number of drops have been added, the mixture in the well will have an orange tint. This first appearance of a uniform orange tint signals the end point. Make a note of how many drops were used. Then add 1 or 2 more drops of silver nitrate with stirring. The color will likely become darker orange or red due to formation of more Ag_2CrO_4.

4. Finally, add more tap water, with stirring, until the color reverts back to a milky yellow-green. Save this mixture for color comparison.

IV. Analyze Tap Water

In doing titrations, your goal is to catch the point where one drop of $AgNO_3$ produces the first permanent color change. It is not a dramatic color change but rather a slight orange tint. This is the "end point" where all of the chloride has been used up by reaction with the $AgNO_3$.

1. Carefully add 20 drops of tap water (count them) to each of four wells (B1, B2, etc.). Make sure that the drops fall directly to the bottom of the well and are not "trapped" along the side. If you make a mistake, simply start again in another well. Record the position numbers (B1, B2, etc.) for the wells you are using on your data sheet.

    Hints for good pipet technique: (1) Best results are obtained when the pipet is held vertically while dispensing drops. If you hold it at an angle, try to always use approximately the same angle. (2) It is useful to set aside one well in the wellplate as a "waste container." If there is an air bubble or a partial drop forming, it can be discarded in the waste container.

2. Add 1 drop of indicator to each well.

3. Start adding $AgNO_3$, <u>one drop at a time</u>, to the first well as you stir the solution. Stop when one drop of $AgNO_3$ produces a permanent orange color. Record the number of drops you used on the data sheet. If you make a mistake, skip this trial and go on to the next one.

    Hint: The gradual color change from yellow to pale orange may be hard to recognize. If so, it is helpful to use the mixture in well A1 for comparison. You are looking for the first change away from that color.

4. Repeat this titration for each of the other 3 wells. With a little practice, the titrations go very fast. Record the number of drops you used for each well on the data sheet.

    Hint: Once you know approximately how many drops of $AgNO_3$ are needed, you can quickly add about 2/3 of that amount. For example, if the first titration required 15 drops of $AgNO_3$, then for the next trial, you could quickly add about 10 drops of $AgNO_3$ and then add the remaining drops slowly.

    Note: If you missed the end point or lost count of drops or are uncertain of the result for any other reason, it is good scientific practice **not** to erase the data. Simply make a note in the table as to what you think went wrong, draw a diagonal line through that column, and do another titration. All scientists make mistakes when doing lab work! Results crossed out will not affect your grade for the experiment.

5. If desired, titrate an additional one or two samples of tap water.

V. Analyze Collected Water Samples

1. Rinse the pipet and beaker used for tap water, first with distilled or deionized water and then with a new sample to be analyzed.

2. Do a trial titration with 20 drops of the new sample in one well (e.g., B1). Add 1 drop of sodium chromate indicator, then titrate with silver nitrate as before. Depending on the nature of the sample, it may have very little chloride, requiring only a few drops of silver nitrate, or the chloride concentration may be very high, requiring a large quantity of silver nitrate. Record the number of drops used.

3. Depending on the results from this first titration, you may decide to use more than 20 drops or less than 20 drops of water. Dispense samples of the desired size into three more wells.

4. **Optional extension**: If time permits, you may be asked to analyze one or more additional water samples. (Once you learn how to do them, these titrations go very rapidly.) If additional samples are done, record your results on a separate sheet of paper and attach it to the data sheet.

VI. Clean Up

1. Dump the contents of the wellplate into a designated waste container.

2. Rinse the wellplate with tap water, then with distilled or deionized water if available. Leave it upside down on a paper towel to drain.

3. Your instructor will indicate whether the pipets are to be discarded or saved.

VII. Calculations

You should now have at least four results for tap water and four results for at least one other water sample. The same calculation procedure is used for both sets. The logic of the calculaions is explained on the next page. (For a more detailed explanation of the calculations for a titration analysis, you should refer back to Experiment 13, *Analysis of Vinegar*.)

The chemical equation for the reaction of chloride ion with $AgNO_3$,

$$Cl^- \text{ (in water)} + AgNO_3(aq) \rightarrow AgCl(s) + NO_3^-(aq) \qquad (1)$$

shows that 1 mole of chloride reacts with 1 mole of $AgNO_3$. Thus, for the titration of a water sample, the number of moles of silver nitrate added equals the number of moles of chloride ions (in the water sample) that was in the well.

$$\text{moles of chloride in the well} = \text{moles of } AgNO_3 \text{ added} \qquad (2)$$

For the left side of equation (2), the moles of chloride $=$ (molarity of chloride) $\times$ (vol. of water), and for the right side, the moles of $AgNO_3$ $=$ (molarity of $AgNO_3$) $\times$ (vol. of $AgNO_3$).

Substituting into equation (2) gives

$$(\text{molarity of chloride}) \times (\text{vol. of water}) = (\text{molarity of } AgNO_3) \times (\text{vol. of } AgNO_3) \qquad (3)$$

If we assume that the drops are all the same volume, then we can say that

$$(\text{molarity of chloride}) \times (\text{drops used}) = (\text{molarity of } AgNO_3) \times (\text{drops used}) \qquad (4)$$

Rearranging equation (4) gives the final working equation:

$$\text{molarity of chloride} = \text{molarity of } AgNO_3 \times \frac{\text{no. of drops of } AgNO_3}{\text{no. of drops of water sample}} \qquad (5)$$

1. To do the calculations, first look carefully at the four titration volumes for tap water. Does any one result seem way out of line from the others? If so, you probably did something wrong, and it is legitimate to omit that result. Simply write "omit" underneath it. Using all of the results that you think are valid, calculate the average number of drops of $AgNO_3$ used to titrate the tap water and record this on the data sheet.

2. Use equation (5) to calculate the molarity of chloride in the tap water.

3. Finally, convert the answer to milligrams of chloride per liter of water (mg/L), which is the same as parts per million. To do this, multiply the molarity of chloride by 35,500 mg of Cl per 1 mole of Cl. (One mole of chloride has a mass of 35.5 g, which is the same as 35,500 mg.)

> **Example**: Suppose that you found the chloride concentration was 0.0012 mole/liter. Then the calculations would be as follows:

$$\frac{0.0012 \text{ mole chloride}}{1 \text{ L water}} \times \frac{35,500 \text{ mg chloride}}{1 \text{ mole chloride}} = \frac{43 \text{ mg chloride}}{1 \text{ L water}} = 43 \text{ ppm} \qquad \textbf{(6)}$$

4. Repeat the calculations in steps 1–3 for your collected water sample(s).

Report your results in whatever manner is specified by your instructor.

VIII. Interpretation of the Class Data

When class data are assembled, the class will examine the results to see what patterns emerge. In particular, you will try to assess how much effect the human use of water has on water quality in the local environment. In many places, the water is used by a city and passed on to other communities downstream. It is useful to ask what has been done to it in the process. Finally, your class will try to answer the question, "Where does the extra chloride come from?"

> ### Questions To Be Answered After Completing This Experiment
> Write out answers to the following questions on a separate sheet of paper and hand it in along with the data sheet.

1. During the experiment, you were cautioned about making sure that all drops of solutions were at the bottom of the well, not along the sides. Why?

2. Suppose that during the titrations you consistently added sufficient $AgNO_3$ to change the color of the indicator to a darker orange-red. Would this change the results? Explain.

3. What is the advantage of doing three or more trials, especially since you are doing exactly the same thing each time?

4. After examining the assembled class data, what can you conclude about the impact of your city on the local river or stream?

5. List some of the major nonindustrial sources of chloride in natural waters. Which ones do you think are important in your area?

6. Based on what you have learned in this experiment, do you consider chloride a serious water pollutant? Why or why not?

Name _____   Date _____

Partner _____   Lab Section _____

Data Sheet—Experiment 15

Molarity of the $AgNO_3$ solution: _____ mole/liter

Part IV: Tap Water

Trial #	1	2	3	4	5	6
Well number						
Drops of water						
Drops of $AgNO_3$						
Comments						

(Extra columns are provided in the table in case they are needed.)

Average number of drops of $AgNO_3$: _____ (per 20 drops of tap water)

Concentration of chloride: _____ mole/liter   **(eq. 5)**

Concentration of chloride: _____ mg/liter (or ppm)   **(eq. 6)**

Part V: Collected Water Sample

Description of the water sample: _____

Trial #	1	2	3	4	5	6
Well number						
Drops of water						
Drops of $AgNO_3$						
Comments						

(Extra columns are provided in the table in case they are needed.)

Average number of drops of $AgNO_3$: _____

Concentration of chloride: _____ mole/liter   **(eq. 5)**

Concentration of chloride: _____ mg/liter (or ppm)   **(eq. 6)**

(If additional samples are analyzed, the data can be recorded on the reverse side of this sheet.)

Analysis of Bottled Water

Introduction

Chapter 5 in *Chemistry in Context,* fourth edition, has an extensive discussion of bottled water. In this experiment, you will analyze a sample of bottled water to measure the concentrations of several of the dissolved substances likely to be present in the water. You will use techniques that you may have used in previous experiments, notably titrations with colored indicators and use of a pH meter.

This study will be a class project, involving several different analyses and a variety of bottled water samples. Teams of 4 or more students will take one bottled-water sample and divide up the work for the analyses. At the end, class results will be assembled for comparison and discussion.

Background Information

The sources and composition of bottled water are discussed in *Chemistry in Context*, Chapter 5. Bottled water is rarely, if ever, pure water. Because it often comes from wells or springs, bottled-water frequently has a relatively high mineral content. In particular, it is likely to have calcium and magnesium ions, the primary sources of **water hardness** (see the text, Section 5.15.1). If calcium and magnesium are high, there is usually also a high concentration of bicarbonate ion, HCO_3^-, which accounts for the **alkalinity** or **acid neutralizing capacity** of the water. (Section 6.11 in *Chemistry in Context* discusses acid neutralizing capacity.) In addition, there will be varying amounts of other soluble ionic substances. The text includes a label from one brand of bottled water showing the concentrations of the major substances present.

The pH of water must be close to neutral (pH 7) in order for the water to be safe for human consumption, but it will be slightly on the alkaline side (i.e., above pH 7) if there is substantial bicarbonate ion present.

Another way to characterize bottled-water samples (or any other water samples) is the total amount of dissolved material, often referred to as *total dissolved solids (TDS)*. This consists mostly of ionic substances, since water is an excellent solvent for ions. Some bottled-water labels indicate the TDS concentration.

For easy comparisons, all experimental results (except pH) will be converted to parts per million (milligrams of a substance per liter of water).

Overview of the Experiment (Not all steps may be done)

1. Measure the pH of the water.
2. Measure the concentration of calcium and magnesium by titration.
3. Measure the bicarbonate ion concentration by titration.
4. Measure the chloride ion concentration by titration.
5. Measure the total dissolved solids in the water.
6. Assemble the analytical information and compare results for different samples.

Materials Needed

Chemicals	Equipment
• calibration solutions, pH 7.00 and 4.00	• 24-well plastic culture plate
• EDTA solution, 0.0100 M	• plastic transfer pipets
• buffer solution, pH 10	• pH meter and combination electrode
• calmagite indicator solution	• plastic stirring rods
• silver nitrate solution, 0.001 M	• 2 50-mL beakers
• potassium chromate indicator solution	• graduated cylinder, 25 mL or 50 mL
• hydrochloric acid, 0.0200 M	• balance capable of reading to 0.001 g
• methyl orange indicator solution	• hot plate

EXPERIMENTAL PROCEDURE

This experiment includes several different analyses. Your instructor will specify which part(s) you should do and how the class will be divided.

I. Measurement of the pH

This should be done following the procedure described in Experiment 18, *pH Measurements of Common Subtances*. If you are already familiar with pH measurements and if a pH meter has already been calibrated, the actual measurement will take only a few minutes. If you are doing pH measurements for the first time, you need to read the background information and follow the step-by-step instructions carefully.

An alternate, fast, but less precise, way to measure pH is to use pH test strips.

II. Concentration of Calcium and Magnesium

This can be done by a titration procedure described in Experiment 14, *Measurment of Water Hardness*. Water samples are titrated with a solution of EDTA. The color change of an indicator (calmagite) shows when enough EDTA has been added to exactly react with all of the calcium and magnesium.

> **Note:** If an EDTA solution of known concentration is available, there is no need to do the initial titrations of a reference solution as specified in Experiment 12. In that case, a specific number of drops of water should be used rather than 1 mL. The calculation of calcium + magnesium molarity then becomes

$$\text{molarity of Ca + Mg} = \text{molarity of EDTA} \times \frac{\text{no. of drops of EDTA}}{\text{no. of drops of water}} \qquad \textbf{(1)}$$

For an explanation of the logic behind this calculation, refer to Experiment 13, *Analysis of Vinegar* or Experiment 15, *Measurement of Chloride in River Water*.

Calculations: The analysis measured the *combined* concentration of Ca and Mg. Since calcium is almost always present in much larger concentration than magnesium, we will simplify the calculation by assuming it is all calcium. Calculate the concentration as parts per million, ppm, by one of the following methods: (a) If you calculated the molarity of Ca + Mg by equation (1) above, multiply this molarity by 40,000 mg/mole, the molar mass of Ca expressed as milligrams. (b) If you used the calculation procedure in Experiment 14, multiply the "hardness" by 0.40 to convert from mg $CaCO_3$ per L to mg Ca per L. (Remember that this is actually Ca + Mg.)

III. Concentration of Chloride

This is another titration analysis, described in detail in Experiment 15, *Measurement of Chloride in River Water*. Samples of water are titrated with a silver nitrate solution of known concentration. An indicator (sodium chromate) is added to show when enough silver nitrate has been added to exactly react with all of the chloride.

Calculations: Do these exactly as specified in Experiment 15, reporting the final result as ppm of chloride in water.

IV. Concentration of Bicarbonate

Analysis for bicarbonate is not described elsewhere in this laboratory manual, so more detailed instructions are provided here. The procedure is straightforward, using the same basic titration technique employed in Experiments 13, 14, and 15. In this case, bicarbonate ion (HCO_3^-) in the water is titrated with hydrochloric acid solution (HCl). The equation for the chemical reaction is

$$HCO_3^- \text{ (aq)} + HCl \text{ (aq)} \longrightarrow H_2O + CO_2 \text{ (g)} + Cl^- \text{ (aq)}$$

A colored indicator, methyl orange, will change color from yellow to orange when an excess of HCl is present so that the solution becomes acidic. (More precisely, the indicator changes color when the pH drops below 4.)

A. Get Organized for Titrations

1. Obtain a plastic wellplate with 24 wells and a small stirring rod. Place the wellplate on a sheet of white paper so that color changes can be seen easily.

2. Using a clean, *dry* beaker or other container, obtain about 10 mL of hydrochloric acid solution (HCl). Label the container and record on the data sheet the exact concentration given on the label of the supply bottle.

3. Pour some of the bottled water into another clean, *dry* beaker or other container and label it.

4. Obtain two graduated-stem plastic pipets (or other matched-style pipets). Label one "water sample" and the other "HCl" and place them in the corresponding solution. *Labeling is crucial because it is easy to get them mixed up!*

5. If you have *not* previously done titrations with plastic disposable pipets, you should practice dispensing drops into the wellpate (e.g., somewhere in row D) until you can reliably add one drop at a time. Some useful advice on pipet technique is given in Experiment 13 under "Pipet Practice."

B. Titration of Bottled Water

1. Using the "water sample" pipet, add 10 drops of this water to well A1 in the wellplate.

2. Add 1 drop of methyl orange indicator. The mixture should be yellow.

3. Start adding HCl solution, <u>one drop at a time</u> (counting the drops), with gentle stirring after each addition. Carefully observe what happens. Watch for the *first color change* from yellow to orange that persists after the mixture is stirred. Record the number of drops used.

4. To aid in recognizing the color change for subsequent titrations, add a few more drops of bottled water to well A1. The color should go back to yellow. Save this for color comparison. *You will be looking for the first color change away from this yellow.*

5. From the result of the first titration, decide how much water to use for subsequent titrations so as to require a convenient amount of HCl. If possible, you should aim for at least 10 drops and no more than 30 drops of HCl.

6. Carefully add the desired number of drops of water to wells A2 through A4.

7. Add 1 or 2 drops of methyl orange indicator to each.

8. Titrate the sample in well A2, adding HCl, one drop at a time, to the first color change that persists after stirring. (Use well A1 as a reference color.) Record the number of drops in the data table.

9. Titrate the next two samples in the same manner. If the 3 or 4 results do not show adequate consistency, do one or two additional titrations.

Calculations: The logic of the calculations is the same as that shown in Experiment 15 for chloride ion. First, calculate the molarity of bicarbonate ion by the following equation:

$$\text{molarity of HCO}_3^- = \text{molarity of HCl} \times \frac{\text{no. of drops of HCl}}{\text{no. of drops of water sample}} \qquad (2)$$

To convert to ppm, multiply the molarity of HCO_3^- by 61,000 mg HCO_3^- per mole of HCO_3^-.

V. Concentration of Total Dissolved Solids

1. Check to be sure that a large hot plate has been turned on and is ready to use.

2. Obtain two clean, dry 50-mL beakers. Label them with numbers and your initials. Weigh each and record the mass to the nearest milligram (0.001 gram). *First check to be sure the balance reads 0.000 g with nothing on the pan.*

3. Using a graduated cylinder, carefully measure out 20 mL of bottled water and add it to the weighed beakers.

4. Place the beakers on a hot plate, adjusted so that water will boil gently. (Be sure you can identify your sample since other samples will probably be on the same hot plate.) Continue heating until the water has *completely* evaporated from the beakers.

5. Remove the beakers (*caution: use tongs*) and allow them to cool. Inspect carefully to be sure they are totally dry, with no hint of moisture remaining in the bottom or on the sides of the beakers.

6. Reweigh the beakers and record the masses. *First check to be sure the balance reads 0.000 g with nothing on the pan.*

<u>Calculations</u>: Subtract to find the mass of solid in a beaker, then calculate the ppm of TDS as follows. First convert the volume of water used to liters (e.g., 20 mL is 0.020 L).

$$\text{ppm of TDS} = \frac{\text{g of solid x 1,000 mg/g}}{\text{vol of water (in L)}} \tag{3}$$

VI. (Optional) Use an Instrumental Procedure to Measure Calcium and Magnesium.

The titration procedure in Part II is a fast and relatively simple way to measure the *combined* concentrations of calcium and magnesium. A more elegant method can be used to measure each ion separately. This requires an expensive, partially automated instrument that is typical of equipment used in modern chemical analysis laboratories. If this step is assigned, your instructor will provide detailed instructions for the particular equipment being used. The general strategy is to measure absorption or emission of radiation for a series of standard reference solutions of each metal, then measure the water sample and compare it with the standards. A calibration graph is needed, which is either prepared by hand or done electronically by a computer within the instrument. (An example of such a calibration graph is shown in the text, Figure 5.24 in the fourth edition.) One advantage of this kind of analysis is that, once the calibration has been done, many samples can be analyzed very rapidly.

Clean Up

When finished with all analyses, your instructor will tell you whether the solutions should be put into a designated waste container or whether they can be poured down the drain (since they are all quite dilute). Your instructor will tell you what to do with the plastic pipets. The remaining glass and plastic items should be rinsed thoroughly with deionized or distilled water and left upside down to drain.

Comparison of Results

Your instructor will specify how this is to be done. Your team should first assemble the analyses for your particular water sample, then post them or report them in some fashion to the rest of the class. Do different bottled brands show significant differences? In some cases, it may be possible to compare student results with data printed on the labels.

Reporting Your Results

Your instructor will specify what kind of report should be submitted.

Questions To Be Answered After Completing This Experiment

Write out answers to the following questions on a separate sheet of paper and hand it in along with the data sheet.

1. Two negatively charged ions were measured, chloride and bicarbonate. Which is present in higher concentration in the bottled-water sample? To answer this question from a chemical viewpoint, you will need to compare *molarities*.

2. Using your results, what is the ratio of the bicarbonate *molarity* to the combined *molarity* of calcium + magnesium? If the calcium/magnesium and bicarbonate all came from limestone dissolved in the water, then the bicarbonate molarity should be exactly twice that of the calcium + magnesium molarity (Equation 6.26 in the text). How closely do your results agree with this prediction?

3. The pH of water for human consumption should be close to neutral. How closely does your bottled-water sample compare with this goal?

4. Check the labels on the bottled waters that you studied. Were they merely treated, or were they purified in some way? If so, list the method(s) used. What was the specific goal of the treatment? See Section 5.16 of your textbook for help.

5. **Optional** (This question is more challenging.) The text shows a label from one brand of bottled water. For the positive ions that are listed, calculate the total *molarity* of all positive ion charges. (Keep in mind that an ion such as calcium, Ca^{+2}, has two positive charges.) Do the same thing for the negative ions. (If sulfate, SO_4^{2-}, is present, it has two negative charges.) How well does the *sum* of the positive charge molarity equal the *sum* of the negative charge molarity?

Name _____   Date _____

Partner/team members _____   Lab Section _____

Data Sheet—Experiment 16

Note: For the parts done by other members of your team, note this and simply enter the final results.

Description of bottle-water sample _____

Part I: pH _____

Part II: Calcium and Magnesium

(Use the data table in Experiment 14 or the following table)

Trial #	1	2	3	4	5
Well number					
Drops of water					
Drops of EDTA					
Molarity of Ca + Mg (equation 1)					

(extra columns are provided if needed)

Average molarity: _____ mole/liter

Calculated ppm of Ca + Mg _____ mg/L

Part III: Chloride

Molarity of the AgNO$_3$ solution: _____ mole/liter

Trial #	1	2	3	4	5
Well number					
Drops of water					
Drops of AgNO$_3$					
Molarity of Cl (eq 5, Experiment 15)					

(extra columns are provided if needed)

Average molarity: _____ mole/liter

Calculated ppm of chloride _____ mg/L

Name _____ Date _____

Partner/team members _____ Lab Section _____

Data Sheet—Experiment 16, page 2

Part IV: Bicarbonate

Molarity of the HCl solution: _____ mole/liter

Trial #	1	2	3	4	5
Well number					
Drops of water					
Drops of HCl					
Molarity of HCO_3^- (equation 2)					

(extra columns are provided if needed)

Average molarity: _____ mole/liter

Calculated ppm of HCO_3^- _____ mg/L

Part V: Total Dissolved Solids

	Trial 1	Trial 2
Volume of water used, mL		
Mass of beaker, g		
Mass of beaker + solid, g		
Mass of solid, g		
TDS, mg/L (equation 3)		

Reactions of Acids with Common Substances

Introduction

Acids are an important category of chemical substances. They are found in many places, including certain foods, our bodies, and frequently in rain or snow. (See Chapter 6 in *Chemistry in Context* for more detail.) Acids undergo chemical reactions with a great variety of substances. Some of these reactions are desirable, while others can be quite damaging. The acids found in rain or snow can have damaging effects on various materials.

In this experiment, you will investigate the reactions of three common acids (hydrochloric acid, sulfuric acid, and nitric acid). You will study their reactions with some familiar substances: marble (a building material), four common metals (zinc, copper, nickel, and aluminum), and a sample of protein (egg white) that represents living matters. In order to make observations in a very limited time, you will use acid solutions that are far more concentrated than those found in rain or snow. Nevertheless, the same reactions do occur with acid rain, but on a vastly slower time scale.

Background Information

Acids are substances that contain hydrogen ions (H^+) or that react with water to form hydrogen ions. Chapter 6 in *Chemistry in Context* provides more detail on acids and discusses the environmental consequences of the acids found in rain.

Most metals react with acids to produce hydrogen (H_2), a colorless gas, and metal ions. However, as you will see, not all acids and metals react in exactly the same way. Careful observation will also show that the rate of reaction of acids with materials varies with concentration. Building materials vary widely in their reactivity with acid. Some, like marble (a form of calcium carbonate), react readily with acids. Others, such as granite, do not visibly react at all. The reaction of proteins with acids is unlike that of any of the other materials. In aqueous solutions, protein molecules have a preferred three-dimensional shape that is pH dependent. Addition of an acid changes the pH and the protein shape so that the protein becomes less soluble. Again, reactions vary. In addition to causing solubility changes, some acids also chemically react with the protein molecule itself.

In this experiment, the concentrations of acids are expressed as molarity, which is discussed in *Chemistry in Context,* fourth edition, Section 5.6. The **molarity** of a substance in solution (often designated by M) is defined as the number of moles of that substance in 1 liter of solution.

Overview of the Experiment

1. Test the reactions of three acids (HCl, H_2SO_4, HNO_3) with marble chips (calcium carbonate).
2. Test the reactions of these acids with several metals (zinc, copper, nickel, aluminum).
3. Test the effect of these acids on egg white (a sample of protein).
4. Find out whether a change in acid concentration affects the speed of the reactions.

Materials Needed

Chemicals	Equipment
• 6 molar hydrochloric acid • 6 molar sulfuric acid • 6 molar nitric acid • 0.6 molar sulfuric acid • 4 small pieces of marble • 4 small strips of zinc or galvanized nails • 4 paper clips (new and shiny) • 4 small strips of aluminum • egg white	• plastic wellplate, 24 wells • disposable plastic pipets, small size • wash bottle with distilled/deionized water

SAFETY NOTES

• You should always wear eye protection in the laboratory, but this is especially important when working with acids. Safety glasses are absolutely essential to protect your eyes in case of any splashes.

• Be careful not to spill acids on your clothing, books, papers, or yourself. If you do spill any acid on your skin, wash it off promptly with large amounts of water. Continue to run water on the affected area for several minutes. (You should also check with your instructor to see if further attention is needed.) If you spill any acid on the lab bench, carefully wipe it up with a wet paper towel and then rinse the area with a second wet paper towel. Rinse the towels with water and discard in a wastebasket.

• Be careful not to get your face too close to the wellplate or to breathe the gases released in the reactions.

EXPERIMENTAL PROCEDURE

I. Preparation of Metal Samples

The tests with metals will work best when the metal surface has been freshly cleaned to remove any corrosion or coating. Therefore, you should obtain the following sets of metal pieces and clean at least one surface of each piece with either sandpaper, emery paper, or steel wool.

1. Zinc: These can be either small, thin strips of zinc or galvanized nails (which are made of iron coated with zinc). When clean, the surface should be bright and shiny.

2. Copper: These may be very short pieces of heavy-gauge copper wire or small pellets of copper shot. Be sure the surface is bright and shiny.

3. <u>Paper clips</u>: Fresh, shiny paper clips should not need any cleaning. These are usually made of iron or steel with a coating of nickel.

4. <u>Aluminum</u>. These can be obtained most conveniently by cutting small strips out of an empty beverage can and then cleaning one surface to expose fresh metal.

II. Preparation of the Wellplate

1. Place a clean, dry 24-well wellplate on a sheet of white paper. Label the rows and columns on the paper according to the diagram below. If possible, also have available a dark surface on which to place the wellplate (e.g., a bench top or dark paper). Some tests show up better when viewed against a dark background rather than a white background.

2. Add one dropper-full (about 20–30 drops) of 6 molar hydrochloric acid (HCl) to each of the wells in row A. The wells should be 1/4 to 1/2 filled.

3. Similarly, add one dropper-full of 6 molar sulfuric acid (H_2SO_4) to each well in row B.

4. Row C can be filled by 10-fold dilution of the 6 molar sulfuric acid: e.g., 18 drops of water + 2 drops of 6 molar acid.

5. Fill all wells in row D with 6 molar nitric acid (HNO_3).

6. Have available a paper towel on which to lay test objects after they have been used. Also have available a plastic wash bottle filled with pure water.

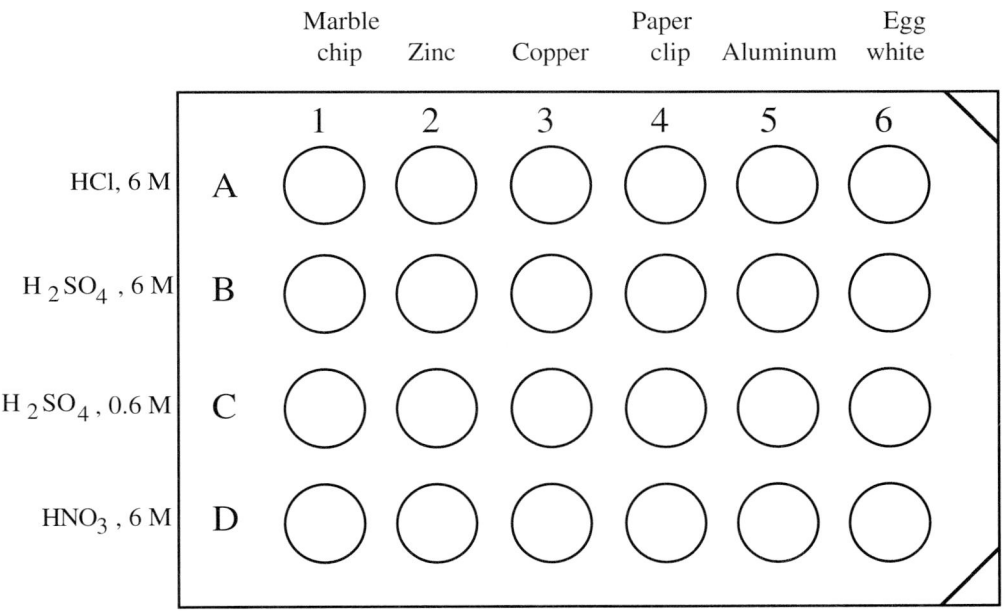

Figure 17.1 Diagram of the wellplate setup

III. Investigation of Reactions

Note: This experiment involves many tests and many observations. Careful observation is essential because the reactions of acids are so varied. *It is important to record your observations as you make them, rather than waiting until the end.*

> **CAUTION:** Nitric acid reacts with most metals to produce poisonous gases. Therefore, all of the tests with metals and nitric acid MUST be done in a fume hood (step 7 below).

1. Put a small marble chip in each well in column 1. Observe carefully what happens and record all of your observations. Do you see evidence of chemical reactions? Pay particular attention to differences between the acids and to differences in reaction rates.

    (**Optional extension:** Test the reaction of these acids with two other familiar forms of calcium carbonate: chalk and eggshell.)

2. Put zinc strips into the hydrochloric and sulfuric acid wells in column 2 (but *not* into the nitric acid well). Try to leave a portion of metal sticking out of the acid so that you have a "handle" with which to remove it when the test is completed. Observe the reaction of the zinc with each acid solution and record your observations. Pay particular attention to differences in the reactions and to differences in reaction rates.

3. Place clean pieces of copper in the hydrochloric and sulfuric acid wells in column 3 (but not into the nitric acid well). Observe the reaction of the copper with each acid solution and record your observations. Pay particular attention to differences in the reactions and to differences in reaction rates.

4. Put small, shiny paper clips into the hydrochloric and sulfuric acid wells in column 4 (but not into the nitric acid well). Observe the reaction of the paper clip with each acid solution and record your observations. Pay particular attention to differences in the reactions and to differences in reaction rates.

5. Put a small piece of aluminum in each well in column 5. Observe the reaction of the aluminum with each acid solution and record your observations. Pay particular attention to differences in the reactions and to differences in reaction rates.

6. Put several drops of egg white (protein) in the wells in column 6. Observe the reaction of the egg white with each acid solution and record your observations. Pay particular attention to differences in the reactions and to differences in reaction rates.

7. Finally, take your wellplate and fresh samples of the four metals to a fume hood. When the wellplate is in the fume hood, put small pieces of zinc, copper, aluminum, and a paper clip into the appropriate nitric acid wells in row D. (*CAUTION: Do not put your face close to the wellplate because a poisonous gas may be produced.*) Observe the reaction of the metals with the nitric acid solution and record your observations. Pay particular attention to differences in the reactions and to differences in reaction rates.

    CAUTION: When finished, either leave your wellplate in the fume hood until the evolution of gas has stopped or else lift out the pieces of metal before removing the wellplate from the hood.

IV. Clean Up

When you have completed the experiment, empty the contents of your wellplate into the waste container provided. If there are any pieces of marble or metal remaining in your wellplate, put them in the appropriate containers. Finally, rinse the wellplate several times with tap water. Your instructor may give further instructions on what to do with the wellplate.

Questions To Be Answered After Completing This Experiment

Write out answers to the following questions on a separate sheet of paper and hand it in along with the data sheet.

NOTE: This experiment involved many tests and observations. These questions are designed to help in interpreting and summarizing the results of the tests.

1. What general conclusions can you draw from your observations? Do acids react with these materials? Did the reactions all appear to be the same?   If there were differences, was there any pattern to the behavior?

2. Marble is calcium carbonate, $CaCO_3$. Its reaction with hydrochloric acid is

$$CaCO_3 + 2\,HCl \longrightarrow CaCl_2 + CO_2 + H_2O$$

What <u>gas</u> was produced? What reactions do you think occurred when marble was exposed to the other acids? Do these reactions have anything in common? In light of your observations, speculate on the damage to marble statues and buildings caused by acid rain.

3. The reactivity of marble with sulfuric acid seems curiously out of line with all of the other observations. If you observed carefully, you probably saw a brief burst of gas bubbles, which quickly stopped. This is because one product of the reaction, calcium sulfate, $CaSO_4$, is very insoluble and provides a protective coating on the marble. This slows the reaction dramatically. Question to ponder: Does this mean that acid rain containing sulfuric acid should have little or no effect on statues and buildings?

4. Do zinc, nickel, iron, and aluminum react with acids?   On the basis of your observations, comment on the effectiveness of zinc coatings and nickel coatings as a protection for iron against acid rain.

5. Challenge: The gas produced from the reactions of metals with acids was hydrogen, H_2. Write a chemical equation for the reaction of zinc (Zn) with hydrochloric acid. One product is zinc chloride, $ZnCl_2$.

6. Copper has been important since ancient times as a building material used on roofs of buildings. On the basis of your observations, is copper a good material for this purpose? (Note: You may have observed a green color of roofs on public buildings. This is a form of copper carbonate, formed by slow reaction of copper with nitric acid plus the carbon dioxide in the atmosphere.)

7. What generalization can you make about the effect of changing the concentration of an acid? Suppose the acid concentrations were reduced 100-fold, 1,000-fold, or even more?

8. Aluminum reacted differently from the other metals. What differences did you observe? Aluminum easily forms a very unreactive oxide coating. How can this account for the observed differences in aluminum's reactivity?

Notes

Name _____   Date _____

Partner _____   Lab Section _____

Data Sheet—Experiment 17

Acid	Observations with marble
6 M HCl	
6 M H_2SO_4	
0.6 M H_2SO_4	
6 M HNO_3	

Acid	Observations with zinc
6 M HCl	
6 M H_2SO_4	
0.6 M H_2SO_4	
6 M HNO_3	

Acid	Observations with copper
6 M HCl	
6 M H_2SO_4	
0.6 M H_2SO_4	
6 M HNO_3	

Name _____   Date _____

Partner _____   Lab Section _____

Data Sheet—Experiment 17, page 2

Acid	Observations with paper clip
6 M HCl	
6 M H_2SO_4	
0.6 M H_2SO_4	
6 M HNO_3	

Acid	Observations with aluminum
6 M HCl	
6 M H_2SO_4	
0.6 M H_2SO_4	
6 M HNO_3	

Acid	Observations with egg white (protein)
6 M HCl	
6 M H_2SO_4	
0.6 M H_2SO_4	
6 M HNO_3	

pH Measurements of Common Substances

Introduction

pH is a powerful and widely used way of describing the acidity or alkalinity of substances. In this experiment, you will learn how a modern pH meter works, how to calibrate it, and how to make reliable measurements. You will measure the pHs of a variety of substances in water, including some pure chemicals, some rain and water samples, and various foods and household products. You may also have the opportunity to try some simple "dip sticks," which are disposable pH test strips.

Because this experiment includes measuring the pHs of a variety of common foods and household substances, you may wish to bring a few substances of your own choosing to lab so you can check their pHs.

Background Information

The pH scale is a convenient way of expressing hydrogen ion concentration in solutions. Section 6.4 in the text discusses pH and the pH scale. pH is defined as the negative of the logarithm of the hydrogen ion molarity. (See Appendix 3 of the text for more detail.) In equation form, this is

$$pH = -\log M_{H^+}$$

Modern pH meters are wonderfully easy to use and generally reliable. But the science behind their operation is far more complicated. They rely on an *extremely* small electric current flowing through the solution and the meter, and thus they are highly susceptible to a variety of interferences. Probably the most important considerations in making accurate pH measurements are (a) *careful* adjustment and rechecking of the meter, (b) *thorough* rinsing of the electrode with pure water between each measurement, and (c) care in handling the electrode because it is very fragile (and also expensive).

Although the following directions describe the use of an electronic instrument for measurement of pH, there is a much simpler method for measuring approximate pH values using special pH test paper strips. These have been impregnated with colored dyes that change color depending on the pH. Most of this experiment can be done satisfactorily with pH test strips instead of a pH meter. One disadvantage of test strips is that most of them are reliable only to the nearest whole-number pH unit. For some brands of test strips, the pH can be estimated to within a few tenths of a pH unit.

Overview of the Experiment

1. Learn the operation of a pH meter and calibrate the meter.
2. Measure the pH of pure acid and base solutions.
3. Measure the pH of various foods and household products.
4. Measure the pH of tap water, rainwater, and surface water.
5. Measure the pH of water containing two atmospheric gases: CO_2 and (optional) SO_2.

Materials Needed

Chemicals	Equipment
• calibrating solutions: pH 7.00 and 4.00 • HCl solutions: 0.0001, 0.001, 0.01, and 0.1 M • NaOH solution: 0.001 M • foods and household products to be tested • source of carbon dioxide gas • source of sulfur dioxide gas	• pH meter with attached electrode • plastic wash bottle filled with pure water • 2 small beakers • large beaker (waste container) • box of tissues or Kimwipes™ • test-tube rack • supply of 4-inch test tubes (at least 10)

EXPERIMENTAL PROCEDURE

Note: The following directions assume that you will be using a pH meter. If you are using pH test strips instead of a pH meter, you can skip Part I and go directly to the tests in Parts II to V.

I. Learn the Operation of a pH Meter and Calibrate the Meter

1. Obtain a pH meter with an attached probe called a "pH electrode." Your instructor will explain the type of pH meter to be used and arrangements for sharing meters between students. If the pH meter requires AC voltage, be sure it is plugged in and warmed up.

2. Obtain the other equipment items listed.

3. Obtain two small beakers containing calibrating solutions labeled "pH 7.00" and "pH 4.00."

4. Learn how to operate the pH meter, including the functions of whatever knobs or buttons it has. Most meters have three knobs: TEMPERATURE, SLOPE, and CALIBRATE (or STANDARDIZE).

5. Hold the electrode over a waste container and rinse the electrode thoroughly with pure water from the wash bottle. Blot the end of the electrode *gently* with a tissue to remove most of the water.

6. Insert the electrode into the pH 7.00 calibrating solution, stir gently for a few moments, and observe the reading on the meter. Use the CALIBRATE knob to adjust the meter until it displays 7.00. You may need to keep stirring. The meter may not settle down to display *exactly* 7.00, but it should be within the range 6.9 to 7.1.

7. Rinse the electrode thoroughly with pure water, blot the end of the electrode *gently* with a tissue, then insert the electrode into the pH 4.00 calibrating solution, stir and observe until the reading is steady. If necessary, adjust the SLOPE (or TEMP) control until it reads very close to 4.00. (Optional: Check the pH 7.00 again to be sure.)

8. Recommended: Repeat with both pH 7.00 and 4.00 (rinsing and blotting each time the electrode is moved) to be sure the readings are steady.

II. Measure the pH of Chemical Solutions of Known Concentration

1. Remove the electrode and rinse it <u>very thoroughly</u> with pure water, since it is now coated with a concentrated pH 4 or pH 7 solution, which could make subsequent readings erroneous. Blot the end of the electrode *gently* with a tissue to remove most of the water.

2. Immerse the electrode into a 0.0001 M solution of HCl (in a test tube or small beaker), stir the solution for a few moments, and record the pH. Calculate the predicted pH for this solution and record that also.

3. Remove the electrode, rinse it well, blot with a tissue, then immerse the electrode into 0.001 M HCl, stir, and record the pH along with the predicted pH.

4. Rinse the electrode and measure the pH of 0.01 M HCl.

5. Finally, switch to an alkaline (basic) solution, as follows. Rinse the electrode thoroughly and measure the pH of a 0.001 M solution of sodium hydroxide, NaOH. Compare your result with the calculated pH for this solution. (Review how to calculate pH for a base.)

III. pHs of Foods and Household Substances

Two or more student teams can divide up the tests and share the results with each other.

Now that you know how to measure pH, you should measure the pHs of some common substances, using samples provided by the instructor and samples brought by students. In each case, put a <u>small</u> amount in one of the wells of a plastic wellplate or in a small test tube (or in a very small beaker). Use only enough so that the bulb end of the electrode is immersed. If in doubt about this, check with the instructor.

Some possible samples to test include the following substances. You should test 6–8 substances or whatever number your instructor specifies.

a. Vinegar
b. Lemon juice
c. Fruit juices
d. Soft drinks
e. Coffee or tea

<u>The following substances should be diluted before testing the pH. To do this, mix a few drops of the substance (or a small pinch of solid) with pure water.</u>

f. Liquid dish detergent
g. Dishwasher detergent
h. Laundry detergent
i. Shampoo or hand soap
j. Household ammonia
k. Liquid laundry bleach
l. Baking soda (sodium bicarbonate)
m. Drain cleaner (***Caution:*** *see note below*)

> **SAFETY NOTE:** Some drain cleaners are extremely caustic. They are designed to dissolve hair and other debris! Be especially cautious about getting any drain cleaner on your skin. Saftey glasses are essential. *In case of any skin contact, consult the instructor immediately.*

IV. pH of Tap Water, Rain, and Surface Water

1. Measure the pH of tap water. You can try water from the lab, from a drinking fountain, or from elsewhere on campus or where you live.

2. The pH of rain or snow is particularly interesting because of concern about "acid rain," but good measurements are slightly difficult to do. Your instructor may provide samples that have been collected recently or you can set a clean wide-mouth jar outside just before a rain event. Remember that rain is nearly pure water, and this makes the measurements more difficult. The recommended procedure is described in Experiment 19.

3. It is instructive to compare the pH of rain to the pH of river water. If river samples are available in the lab, check the pH of one or more. If they are different from the pH of rain, try to suggest an explanation.

4. If water from a swimming pool is available, check its pH. Is it different from tap water?

V. pH of Water Containing Two Atmospheric Gases: CO_2 and SO_2

1. Test the pH of water that is saturated with carbon dioxide. One easy way to do this is to use seltzer water that has been left open so that the excess CO_2 has escaped and it has gone "flat." Another way is to bubble some CO_2 gas (from a pressurized tank) into water. Yet another way is to use the procedure described in Experiment 1. Test the pH of the water containing CO_2. Is the solution acidic? Is it as acidic as the SO_2 solution?

2. Test the pH of water that is saturated with air. Remember that CO_2 is only a very small fraction of the gases that make up air. How does your measured pH value compare to the value stated in the text?

3. Find out whether sulfur dioxide gas, SO_2, produces acidic solutions as claimed in the text.

> **SAFETY NOTE:** Sulfur dioxide is a very toxic gas. In addition, some individuals have a serious allergic reaction to SO_2. *This test should be done only in a good fume hood and should not be attempted by individuals who are allergic to SO_2.*

To do this test, it is first necessary to produce SO_2 in a Ziplock bag. Fill a plastic pipet with 6 M sulfuric acid and place it in a Ziplock bag. Add about 1–2 grams of sodium sulfite (Na_2SO_3) to the bag. Close the bag securely, making sure to exclude air, and then squeeze the pipet so that the chemicals will mix and react. **Do the next part in a fume hood**. With the bag lying on a table, very cautiously open a corner of the bag, insert a dry plastic pipet (with air squeezed out) and fill it with the SO_2 gas. Reseal the bag immediately. Slowly bubble this gas into pure water in a test tube or wellplate. Then test the pH of the water. Has the pH changed? Is it acidic or basic? When finished, empty the bag in a fume hood.

Clean Up

Leave the pH electrode soaking in water. Clean all glassware, first with plenty of tap water, then finally with pure water. Leave upside down to drain.

Reporting Your Results

Record all of your pH results on the data sheet. If requested by the instructor, post the pHs of common substances for other students to see.

Possible Extension: Use of pH Test Strips for Tests Outside the Lab

Your instructor may give you some plastic pH test strips to take with you so that you can test samples outside of the laboratory, for example, in the dorm, cafeteria, or dining room. If so, make a record of your observations and be prepared to report back to the class.

> **CAUTION**: Any tests done on food or beverage samples MUST be done on small portions (in a separate container) that will NOT be consumed.

Questions To Be Answered After Completing This Experiment

Write out answers to the following questions on a separate sheet of paper and hand it in along with the data sheet.

---

1. Were the pH results for HCl and NaOH solutions approximately what you expected? Explain your reasoning. What general rule can you propose for how the pH should change (up or down and by how much) any time the acid concentration increases by a factor of 10? (For example, when HCl concentration changes from 0.0001 M to 0.001 M to 0.01 M, the acid concentration increases by a factor of 10 each time.)

2. What, if any, conclusions can you draw about the pHs of the different categories of common substances, e.g., (a) pHs of beverages? (b) pHs of soaps and detergents? (c) pHs of substances that normally touch human skin?

3. Explain briefly why pH is a useful way of describing acid and base solutions over a very wide range of concentrations.

4. Why was it necessary to dilute substances f to m before measuring the pH? How do you think the measured pH of the diluted sample compares with the pH of the sample itself?

5. (If Part IV was done) (a) Was the rain acidic or alkaline? Since rain is formed by evaporation and condensation in clouds, why is the pH not that of pure water?   (b) Were tap water and surface water acidic or alkaline? What substances may account for this?

6. (If Part V was done) Were the pHs of CO_2 and SO_2 in water about what you expected? Write chemical equations for the reactions of each gas with water.

Notes

Name _____   Date _____

Partner _____   Lab Section _____

Data Sheet—Experiment 18

Part II:

Solution	Measured pH	Predicted pH
0.0001 M HCl		
0.001 M HCl		
0.01 M HCl		
0.001 M NaOH		

Part III:

Sample, including source or description	Measured pH

Parts IV and V:

Solution or Sample	Source	Measured pH
tap water		
rain		
river or lake		
CO_2 in water		
SO_2 in water		

What Is the pH of Rain?

Introduction

Chapter 6 in *Chemistry in Context* examines the problems associated with acidic rain. pH is the primary way of classifying the acidity of rain samples. Studies of the pH of rain and snow are of great importance as part of national and international efforts to monitor the amounts of sulfur and nitrogen oxides in the atmosphere. In this experiment, you will measure the pH of some rain or snow samples. Although studies of rain pH are easy to do, adherence to correct procedures for collecting and handling the samples is essential for reliable measurements. The directions for this experiment focus on the collection and handling of samples. Your instructor will supply the exact details of the experiment and instructions for using the type of pH meter available in your laboratory.

Background Information

The pH scale is a convenient way of expressing hydrogen ion concentration in solutions. pH is defined as the negative of the logarithm (to base 10) of the hydrogen ion molarity. In equation form this is

$$pH = -\log M_{H^+}$$

An interesting feature of the pH scale makes use of the fact that **all** water samples contain some hydrogen ions. Even alkaline solutions that have excess OH^- still contain some H^+. The product of the molarities of H^+ and OH^- at 25°C is always 1.0×10^{-14}. In pure water, the hydrogen ion concentration will be 1×10^{-7} molar, and the pH will be 7. In acidic solutions, the H^+ ion is in excess, and the concentration will be higher than 1×10^{-7} molar. Therefore, the pH will be lower than 7.

Pure rainwater has a pH of approximately 5.6, due to the presence of carbon dioxide that reacts with water producing carbonic acid that dissociates into hydrogen ions and bicarbonate ions.

$$CO_2(g) + H_2O \rightarrow H_2CO_3 \rightarrow H^+ + HCO_3^-$$

When rain has a pH below 5.6, this means that other acids must be present, and then it is called "acidic rain." Rain can be classified as follows:

Rain classification	pH
Pure rain	5.6
Slightly acidic	5.0 – 5.6
Moderately acidic	4.5 – 5.0
Highly acidic	4.0 – 4.5
Extremely acidic	below 4.0

Overview of the Experiment

1. Prepare sample containers for collecting rain samples.
2. Select sites for the collection containers.
3. Collect rain sample.
4. Prepare the sample for pH measurement.
5. Calibrate the pH meter.
6. Measure the pH of the sample.
7. Analyze the sample data.

Materials Needed

Chemicals	Equipment
• 6 M hydrochloric acid • distilled or deionized water • pH buffer solutions pH 7.00 • pH buffer solutions pH 4.00	• plastic collection bottles • large plastic funnel • pH meter • several small plastic beakers • plastic wash bottle

EXPERIMENTAL PROCEDURE

I. Prepare Sample Containers

Samples should be collected in plastic containers, not glass. It is important that the containers be thoroughly cleaned in advance. The recommended cleaning procedure is to first rinse each container with 6 molar hydrochloric acid. **(Caution: 6 M hydrochloric acid is a corrosive liquid. Wear gloves and a lab apron along with your safety glasses.)** This first cleaning is followed by 5 rinses with tap water and 5 rinses with distilled or deionized water. Once cleaned and dried, the containers should be lightly capped or covered to keep them scrupulously clean.

II. Collect Rain Samples

If the bottles have a large opening, they can simply be set out to collect rain. Alternatively, in order to obtain larger samples, it may be desirable to use a large plastic funnel (also cleaned in advance). In this case, the bottle will probably need to be supported in an upright position—for example, by using an empty metal can nailed to a board. The collecting container should be placed in an unobstructed location where it will not be disturbed. Ideally, it should be at least several feet above ground level in order to minimize contamination of the samples by dust and dirt.

A single sample can be collected for an entire rain event. Alternatively, samples can be collected at intervals during the event. For example, for an afternoon thunderstorm, the container might be changed every 15 minutes. For a one-to-three-day rain event, the container might be changed every few hours or twice a day. The pH of rain can change significantly during the course of a rain event.

III. Handling and Storage of Samples

Bottles should be labeled immediately (date, total sampling time, location) and brought to a laboratory for measurement as soon as possible. It is important that each sample be filtered (into another clean bottle!) to remove any dust and debris that might react with the acids in the rain. Ideally, the samples should be filtered immediately after collection. If this is not possible, they should be kept cold until brought to the laboratory and then filtered as soon as possible. If it is not possible to test the samples within a few hours, they should be stored in a refrigerator.

IV. Calibration of the pH Meter

The measurement of pH is normally done with a pH meter. Two "electrodes" (or a combination of the two in one unit) are attached to a sensitive voltmeter. When they are immersed in a water solution, the voltage between them is proportional to the pH of the solution. A pH meter is convenient and fast to use, but it needs to be calibrated first with solutions of known pH.

1. Set up the pH meter for calibration by following the instructions provided by your instructor or in an instruction manual.

2. Rinse the electrodes with pure water and immerse them in a standard buffer solution of pH 7.00. Stir for 30 seconds, let stand undisturbed for 30 seconds, and then adjust the meter to read 7.00.

3. Rinse the electrodes thoroughly with distilled or deionized water (a plastic wash bottle is convenient for this). Then immerse the electrodes in a second standard buffer of pH 4.00, stir 30 seconds, let stand 30 seconds, and adjust the appropriate knob to read 4.00.

4. Finally, rinse the electrodes thoroughly and repeatedly with pure water to remove all traces of the buffer solutions. The pH meter is now calibrated and ready to use.

V. Measurement of Rain Samples

The official, recommended procedure from the National Institute of Standards and Technology is as follows: Remove drops of water on the pH electrode by blotting gently. Insert the electrodes in a clean beaker containing 10–20 ml of the rainwater sample. Stir or swirl the solution to ensure homogeneous contact with the electrodes (do not use a magnetic stirrer). Allow the solution to settle for approximately 30 seconds. Record the pH after the reading has stabilized.

Carefully measure the pH values for whatever samples are available. Be sure to rinse the electrodes thoroughly between samples. Record the results on the data sheet.

Clean Up

Leave the pH electrode soaking in water. Clean all glassware, first with tap water, then with distilled or deionized water.

> ## Questions To Be Answered After Completing This Experiment
> Write out answers to the following questions on a separate sheet of paper and hand it in along with the data sheet.

1. Describe the collection site(s) for your sample(s).

2. Collect the class data and classify each of the samples according to the acid-rain scale shown at the beginning of the experiment.

3. If several samples from one rain event are available, how did the pH change during the event?

4. If the samples were collected at different locations, are there any pH differences between locations? If so, how can you explain them?

5. Sometimes, rain samples are found to be not acidic at all and instead are alkaline with a pH greater than 7. What might account for this? If you think you have a possible explanation, what tests could be performed by a chemist to support or refute your idea?

Name _____   Date _____

Partner _____   Lab Section _____

Data Sheet—Experiment 19

Sample description	pH

Solubilities: An Investigation

Introduction

This laboratory exercise is a departure from the usual experiment. It is an attempt to simulate the kind of problem solving that takes place in a scientific laboratory. There are no instructions, procedures, or data sheets. There is simply a problem to solve, which requires reasoning skills and the application of knowledge that has been previously acquired.

A team will typically consist of three or four members, one of whom should be designated as the recorder for the group. This person records ideas, data, experimental procedures, etc. Your group might also want to assign other tasks to members of the group, for example, the "go-fer," the experimentalist, the report writer, the team leader. (Your investigation will be more efficient if you take a few minutes at the beginning to be sure each person has a clear role to play.) You and your team members must come up with a solution to the problem and then be prepared to compare your method and your results with those of the other teams in the class.

The Problem

Determine the solubilities of the following compounds in water at room temperature and rank them in order from least soluble to most soluble: calcium sulfate ($CaSO_4$), potassium aluminum sulfate [$KAl(SO_4)_2$], potassium nitrate (KNO_3), ammonium nitrate (NH_4NO_3), copper nitrate [$Cu(NO_3)_2$] and sodium chloride ($NaCl$). All of these compounds, except the copper salt, are likely to be found in drinking water supplies since they occur naturally in rocks and soil or are used as components in fertilizers. Copper nitrate is included because copper salts are often used to reduce algae in ponds and swimming pools.

Solubility is defined as the maximum mass of solid that can dissolve in a given volume of water. It is usually reported as grams of solid per mL of water or grams of solid per 100 mL of water.

In the interest of cost and convenience, you may use no more than 2 grams of each compound. Develop a procedure for carrying out this task and record it and any observations you make while solving this problem. Your grade will not depend on getting a "correct answer," but rather on the quality of the procedure you develop and the care with which you carry out your investigation.

Materials and Equipment

A variety of materials and equipment will be available that you can use to solve this problem. At a minimum, your "lab" should have graduated cylinders, burets, beakers, flasks, plasticware, stirrers, test tubes, plastic pipets, and a balance.

Reporting Your Results

When everyone is finished, your class will assemble to hear a report from each student team about the method used and their results. Therefore, it is important for your team to keep a complete record of everything you do and the numerical data you obtain. In addition to an oral report to the class, your group should prepare a brief (one- to two-page) written report of your investigation. Your instructor will specify what should be included in the report.

Measurement of Radon in Air

Introduction

Radon is a radioactive gas that is a potentially serious indoor health hazard in some geographic areas. In this experiment, you will use a simple radon detector to measure the radon concentration in a location of your own choosing. A sampling period of at least 4 weeks is necessary in order to obtain reliable data.

Background Information

The atomic number of radon is 86, which places it in group 8A, at the far right-hand side of the periodic table. Elements in group 8A are all gases that are chemically unreactive and generally do not form molecules. Radon is an intermediate product in the radioactive decay of uranium-238. (The U-238 decay series is shown in Figure 7.14 in *Chemistry in Context,* fourth edition.) Uranium is widely distributed on the surface of our planet. For example, granite rocks contain about 4 ppm of uranium. Radon-222 (the longest lived isotope of radon) is an alpha-emitter, with a half-life of 3.8 days. Once produced from uranium, radon is drawn upwards through fissures in the rocks and soils. (Because radon is chemically inert, it passes through soil without reacting.) Radon can enter homes through cracks or other holes in the foundation and is typically more serious in houses with basements. The amount of radon found in homes varies from one part of the country to another. National and state maps of predicted radon concentrations in homes can be found at the following EPA website: http://www.epa.gov/iaq/radon/zonemap.html.

In this experiment, you will use a small plastic disk to detect radon. The disk consists of a high-clarity polymer often used in eyeglasses that is known in the trade as "CR-39." The full chemical name for CR-39 is poly[ethylene glycol bis(allyl carbonate)]. As explained in Chapter 9 of *Chemistry in Context*, polymers are large molecules built from smaller units called monomers. In this case, the monomer is $C_{12}H_{18}O_7$. Although $C_{12}H_{18}O_7$ may look like a large molecule, it is far smaller than the polymer made from thousands of this monomer.

Alpha particles (see Table 7.2 in the text) are a type of ionizing radiation. When they penetrate CR-39, they cause damage in the plastic, probably due to disruption of the polymer chain along the path of penetration. Although the damage is not visible to the eye, it can be revealed by treating the disk with sodium hydroxide, NaOH. Sodium hydroxide etches the sample preferentially in the damaged regions. The etched regions show up as "tracks" when viewed under a microscope. The number of tracks in a given area of the plastic disk can be used to estimate the radon level at the sampled location. Although beta and gamma radiation also penetrate the plastic, these rays mostly pass through without causing damage.

Overview of the Experiment

1. Place a detector disk in an undisturbed location for at least 4 weeks.
2. Etch the disk to produce visible alpha tracks.
3. Count the tracks under a microscope.
4. Calculate the concentration of radon in air by comparison with a control disk.

Materials Needed

Chemicals	Equipment
• 6 molar sodium hydroxide • large beaker of boiling water on a hot plate (shared)	• CR-39 detector disk • etch clamp and paper clip • test tube • microscope (shared) • millimeter ruler

EXPERIMENTAL PROCEDURE

I. Air Sampling for Radon

Start the air sampling early in the semester as directed by your instructor. You need to use as long a sampling period as possible, preferably for *at least* a month. Select a location where the detector can remain undisturbed and where you can retrieve the detector prior to the scheduled lab time for this study. (*Plan this carefully, perhaps adding a label to explain what the item is. Janitors or other persons may unknowingly remove the disk when cleaning up an area!*) The class results will be more interesting if students select many different locations, both off-campus and on-campus, inside buildings and outdoors, different floor levels in buildings, different kinds of ventilation or air circulation systems. One possibility is to use your own home, either taking the detector device there yourself or mailing it to a relative with instructions. In some parts of the country, basements of houses are particularly prone to high levels of radon.

You will be given the following materials:

1. A small piece of CR-39.
2. A small plastic cup, preferably with lid.
3. A small rectangle of cardstock (size depends on the cup being used).
4. Piece of "Scotch" tape.
5. Small piece of thin, single-layer toilet paper (to protect the CR-39 from dust).

Proceed as follows:

1. If the cup has a lid, cut a large circular hole in the lid.

2. Remove the plastic film from the side of the CR-39 that is marked.

3. Make a loop of tape with the sticky side out. Stick the side of the CR-39 that still has a plastic film coating onto the tape. Then stick the CR-39 to the middle of the piece of cardstock.

4. Put the cardstock plus CR-39 into the plastic cup, with the CR-39 pointing up and the card bent in an inverted U so that the CR-39 is near the top of the cup. Cover the plastic-cup opening with the toilet paper. Then snap the lid onto the cup. (If the cup does not have a snap-on lid, use a rubber band to hold the toilet paper in place.)

5. Record the start date and time.

6. At the end of the air-sampling period, record the ending date and time. Store the card with the attached disk in a labeled envelope until time for the laboratory work.

Note: It is important that you record accurately the location for the detector disk (describing it in detail) and the start and stop times (dates and time of day).

II. Etching the Disk

Remember to bring your exposed disk to your laboratory class!

1. Remove the disk from the card to which it has been attached. You already peeled off one of the polyethylene films from the disk. Now peel off the other (from the back of the disk). Gently slip the ring of an "etch clamp" over the top of the disk. (*Caution:The disk is fragile and can break easily.*) Then fasten the ring onto a paper clip that has been bent so there is a hook at each end (see *Figure 21.1*).

Figure 21.1

2. Hook the disk inside a test tube and add enough 6 molar sodium hydroxide solution (NaOH) to cover the disk in the tube. *CAUTION: The coating on the paper clip will react with NaOH; therefore, try not to submerge any more of the paper clip than is necessary.*

> **SAFETY NOTE**: Hot 6 molar NaOH is extremely corrosive to skin and to clothing. *Wear goggles at all times. Rubber gloves are strongly recommended.*

3. Start a beaker of water heating on a hot plate and bring to a boil. Heat the test tube containing the disk and NaOH solution in this boiling water bath for 40 minutes. At the end of 40 minutes, remove the disk and rinse thoroughly with lots of tap water.

4. **Clean up**: When finished, do NOT pour the NaOH solution down the drain. Pour it into an approved container provided by the instructor.

III. Counting the Alpha Tracks

This will be done with a microscope. Your instructor or teaching assistant will show you how to operate the particular microscope that you will be using.

1. Determine the area (in square centimeters) of the field of view of your microscope at low power (10x) by looking through the microscope at a ruler. (Consult your instructor if you are unsure about how to calculate this.) To do this, count the number of millimeter divisions on the ruler that you can see and record this. This is the diameter in millimeters. Divide by 10 to find the diameter in centimeters. The radius, r, is half of this, and the area, A, can be calculated by the formula $A = \pi r^2$.

2. <u>General instructions for counting alpha tracks</u>. Place an etched disk on a microscope slide under the microscope. Then adjust the focus until you are sure you are focused on the etched disk. Since the glass slide and disk are at different heights, they will come into focus at different times as you adjust the focus knob on the microscope. As a result, it is easy to mistake the microscope slide for the disk. To prevent this from occurring, move the microscope slide so that the edge of the disk is in view. Then adjust the focus until the disk comes into sharp focus. Once you have focused on the disk, you can move the slide around to observe the tracks in different areas.

3. Practice looking at alpha tracks on a previously exposed and etched disk. This may be a disk donated by a student in a previous class, or it may be one prepared by your instructor using a radioactive source (in which case, your own disk will likely have fewer tracks).

    Look at the two pictures of alpha tracks at different magnifications (*Figure 21.2*). The right-hand photo (*Figure 21.2B*) is more nearly what you will see. The various shapes of the tracks depend on how the alpha particles entered the solid (see *Figure 21.2A*). The circular-shaped tracks are due to alpha particles that entered straight (perpendicular to the disk), and the more teardrop-shaped tracks are due to alpha particles that entered at an angle.

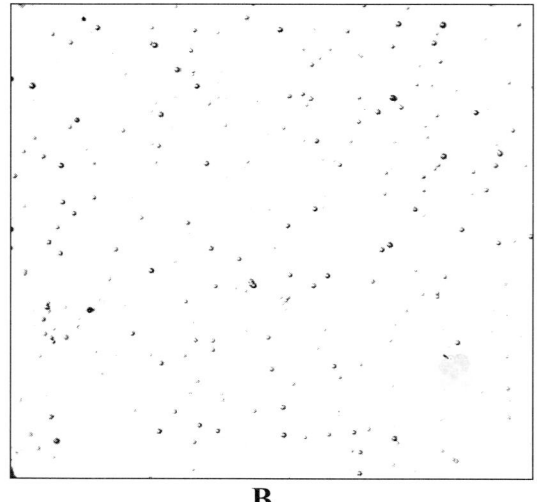

A   B

Figure 21.2 Appearance of alpha tracks on an etched CR-39 disk. (A) At high magnification, showing different shapes of tracks. (B) 20X magnification, sample exposed to 9 pCi/L for five days.

4. Place your own disk on a microscope slide; look through the microscope and adjust the focus until you are sure you are focused on the etched disk. Again, be sure not to mistake the microscope slide for the disk. Once you have focused on the disk, you can move the slide around to observe the tracks in different areas.

5. Count the number of tracks *in 10 different fields of view* and record them on the data sheet.

IV. Calculations

1. Determine the average number of tracks in a field of view for your disk.

2. Then determine the average number of tracks per square centimeter per day of exposure (including any fraction of a day). See your instructor if you need help with this calculation.

3. Control plastic disks that were sent to a radon facility in which the radon level was known to be 370 picocuries per liter were found to exhibit 2,370 tracks/cm^2/day by this etching and counting technique. Given this relationship between picocuries per liter and tracks per square centimeter per day, calculate the radon level in picocuries per liter of air for *your* sample and enter this on the data sheet. How does your sample compare with the 4-pCi/L guidelines set by the Environmental Protection Agency?

Note: Concentrations greater than 4 pCi/L indicate the need for a follow-up test. You can do another one yourself or use an EPA-approved radon monitoring service.

Questions To Be Answered After Completing This Experiment

Write out answers to the following questions on a separate sheet of paper and hand it in along with the data sheet.

---

1. For many types of chemical measurement, you need to subtract out a naturally occurring background level of the chemical being measured. In this case, you don't expect to be able to measure any background radiation from alpha particles, so we don't have to subtract out a background. Why don't we expect alpha particle background radiation?

2. You are comparing your measurements to those in picocuries. What does pico mean? What does a curie measure? See your textbook if needed.

3. Answer each of the following three questions, either individually or in a team as assigned by your instructor.

   a. Consider a beachside home in Mayagüez, Puerto Rico, a university town on the west coast of this volcanic Caribbean island. The home is built on a concrete slab and has screen windows. Is radon likely to be a problem for a home-owner in Mayagüez? Why or why not?

   b. Consider a ten-story office building in chilly Duluth, Minnesota. The building is well insulated and sealed against the winter arctic blasts. Is radon likely to be a problem for the occupants of this building? Why or why not?

   c. Consider a two-story farmhouse in rural Nebraska. The basement has a dirt floor with underlying limestone rock. Is radon likely to be a problem for the farmers who have lived there for generations? Why or why not?

4. Here is the health advisory from the surgeon general:

   "Indoor radon gas is a national health problem. Radon causes thousands of deaths each year. Millions of homes have elevated radon levels. Homes should be tested for radon. When elevated levels are confirmed, the problem should be corrected."

Consult your textbook and list 3 things you can do to correct the problem, if a problem exists. Explain briefly how each of the actions you list will help solve the problem.

Name _____   Date _____

Partner _____   Lab Section _____

Data Sheet—Experiment 21

Part I: Air Sampling

Location of detector: _____

Detector disk put in place   Date_____   Time _____

Detector disk removed   Date _____   Time _____

Total days of air sampling (to nearest day)_____

Part III: Calculation of field of view in microscope

# of millimeters visible on the ruler	Diameter of view (cm)	Area of view (cm^2)

Observed number of tracks (10 different fields of view)

Part IV: Calculations

Average number of tracks per field of view	
Number of tracks per cm^2	
Number of days sampled	
Number of tracks per cm^2 per day	
Radon concentration, pCi/L	

Chemical Reactions and Electricity

Introduction

This experiment is designed to accompany Chapter 8 in *Chemistry in Context.* You will construct several simple electrochemical cells in which chemical reactions produce electricity. You will also investigate the opposite process in which electricity is used to produce chemical reactions.

Background Information

Any spontaneous chemical reaction involving transfer of electrons from one atom or molecule to another atom or molecule can be used as the basis for an electrochemical cell. Consider for example the reaction between zinc metal and copper ions in water solution:

$$Zn(s) + Cu^{2+}(aq) \rightarrow Zn^{2+}(aq) + Cu(s)$$

If the reaction is allowed to occur in a single container (i.e., by adding zinc, $Zn(s)$, to a water solution containing copper ions, $Cu^{2+}(aq)$), the blue color of the copper ions disappears, the zinc gets used up, and the reddish color of metallic copper appears. Electron transfer has occurred, with each zinc atom losing two electrons (to form the +2 ions) and each copper ion gaining two electrons. As explained in the text, we can break this up into two half reactions:

$$Zn(s) \rightarrow Zn^{2+}(aq) + 2\ e^-$$

$$Cu^{2+}(aq) + 2\ e^- \rightarrow Cu(s)$$

In order to get useful electric energy from this reaction, the two half reactions have to occur in two separate locations but connected in such a way that the electrons will flow through an external wire. The wire can be connected to something (such as a motor, lightbulb, or meter) to show that an electrical current is produced. Such a device is correctly called a **galvanic cell** (sometimes called a voltaic cell). The nonscientific and slightly incorrect name is a battery. You will investigate how to assemble several working electrochemical cells, and you will use a voltmeter to measure the voltages produced. **Voltage** can be thought of as the "pressure" driving the electrons from one chemical to the other. This depends on how different the two substances are in their tendency to attract or release electrons.

Rather than allowing a chemical reaction to produce electric energy (in the form of a flowing electron current), the reverse is also possible. In an **electrolytic cell**, electric energy from some source (e.g., a commercial "battery") is used to force a chemical reaction to go backwards, i.e., in the nonspontaneous direction.

Both types of cells have many practical applications. Everyone is familiar with electrochemical cells ("batteries") ranging in size from the tiny cells in hearing aids or watches to large automobile or truck batteries. Electrolysis cells are widely used in chemical manufacturing to produce chemicals that would be otherwise difficult to make.

Overview of the Experiment

1. Assemble 3 galvanic cells and measure their voltages.
2. Construct and test a "citrus cell."
3. Assemble a simple electrolysis unit.
4. Carry out the electrolysis of water.
5. Carry out the electrolysis of potassium iodide solution.
6. Interpret the two electrolysis reactions.

Materials Needed

Chemicals	Equipment
• 0.1 molar copper sulfate solution • 0.1 molar zinc sulfate solution • 0.1 molar silver nitrate solution • 1 molar potassium nitrate solution • small piece of copper metal • small piece of zinc metal • small piece of silver metal • phenolphthalein solution • solid potassium nitrate, KNO_3 • solid potassium iodide, KI	• multimeter or voltmeter with 2 test probes • sandpaper to clean the metal pieces • watch glass • filter paper (same size as watch glass) • scissors • 2 9V batteries • 2 snap caps for the batteries • 2 mini alligator clips, red and black • 2 graphite electrodes or prepared pencils

> **SAFETY NOTE:** Silver nitrate will produce a black stain on skin, although it is not harmful. Try to avoid getting this solution on your skin.

EXPERIMENTAL PROCEDURE

Note: This experiment has two parts that can be done independently and in either order. To facilitate efficient use of equipment, your instructor may assign half of the class to start with Part I (galvanic cells) and the other half to start with Part II (electrolytic cells).

I. Galvanic Cells (electricity from chemical reactions)

You will set up three different galvanic cells:

1. $Zn \rightarrow Zn^{2+} + 2\ e^-$ with $2\ Ag^+ + 2\ e^- \rightarrow 2\ Ag$

2. $Zn \rightarrow Zn^{2+} + 2\ e^-$ with $Cu^{2+} + 2\ e^- \rightarrow Cu$

3. $Cu \rightarrow Cu^{2+} + 2\ e^-$ with $2\ Ag^+ + 2\ e^- \rightarrow 2\ Ag$

In order not to waste chemicals, these will set this up at a small scale using a piece of filter paper placed on a watch glass.

1. Obtain a watch glass and a piece of filter paper of the same size. On the filter paper, draw three small circles, labeling them Ag^+, Cu^{2+}, and Zn^{2+} as shown in the figure. Trace a trail from each circle to the center of the paper. Use scissors to cut out wedges of paper between the circles. Place the filter paper on the watch glass.

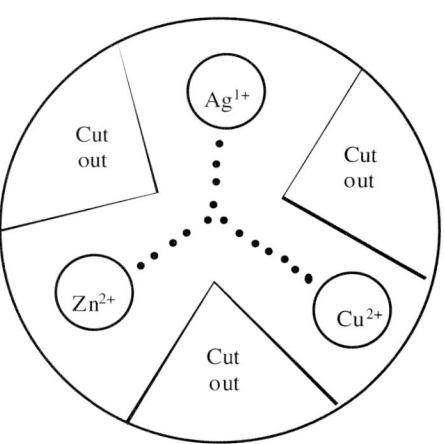

2. Obtain small pieces of each of the three metals: Ag, Cu, and Zn. Using sandpaper, sand both sides of each piece of metal until it is clean and shiny.

3. Place 3 drops each of 0.1 M solutions containing Cu^{2+}, Zn^{2+}, and Ag^+ on the corresponding circles. Then place each piece of metal on the spot of its corresponding ion. *The top sides of the metals should be kept dry.*

4. Now attempt to measure the voltage for the zinc and copper combination as follows. Obtain a voltmeter and learn how it operates. Turn it on and select a range of about 1 or 2 volts. (Some meters automatically select the appropriate range.) Check to be sure that the meter reads zero volts when the two probes are touched together. Then touch the two metal probes from the meter to the pieces of copper and zinc and observe the voltage.

5. It is probable that you observed zero volts (or very close to zero), and you should convince yourself that there is a missing link that prevents a complete cycle or circuit. To complete the circuit, fill a small plastic pipet with 1 molar KNO_3 solution and carefully dribble a small amount of solution along the lines connecting the circles to the middle of the paper. Be sure there is a continuous trail of KNO_3 connecting the circles.

6. Test the voltage again by touching the probes to the Cu and Zn metals. If a positive voltage is displayed, wait 5 seconds and record the reading. If a negative voltage is displayed, reverse the leads, wait 5 seconds, and record the reading. Also record which metal the red probe touches to give a positive voltage.

7. In a similar fashion, measure the voltage for the other two possible combinations: copper + silver and zinc + silver. Record the voltages and which metal the red probe touches to give a positive voltage.

8. The Citrus Cell. The metal ions that were supplied by solution in the last part of this experiment are available from many natural sources, e.g., very hard water or fruits and vegetables. To illustrate this, you can make a "battery" from a piece of citrus fruit. Choose two of the three metals (Zn, Cu, Ag). Push the metal strips through the fruit rind so that a part of each is inside the juicy portion of the fruit. The two pieces of metal must not touch but should be in the same section of the fruit. Record the voltage and which metals you used.

Further Explorations
This study could be expanded to include other metals. Among the possibilities are iron, nickel, tin, and magnesium. Some metals such as aluminum form a very tough coating that makes this kind of measurement impractical. Certain other metals such as sodium and potassium are so reactive that

they will immediately react with the water! Write a brief description of the exploration you plan, check it for safety and feasibility, and, if approved, carry it out.

Clean Up

When done, rinse the small metal pieces with pure water and blot dry with a paper towel. Put the used filter paper in the trash. Rinse off the watch glass with pure water. Return metals and watch glass to the supply location.

9 volts

II. Electrolytic Cells (chemical reactions from electricity)

Assemble a simple electrolysis unit using two 9-volt batteries. (A single 9-volt battery can be used, but two batteries connected in "parallel" are preferable and will give more dramatic results.)

1. Tape two 9-volt batteries together.

2. Obtain two snap-on battery connectors and two alligator clips, one red and one black. Twist the ends of the two red wires together. Thread these combined wires through the hole in the handle of the red alligator clip. Bend the wire back on itself and twist to make it snug. Repeat this process for the black wires.

3. Obtain two graphite rods. (Ordinary lead pencils with a good point work well for this but must have a place to attach an alligator clip, either by sharpening the top end or by cutting away the wood near the middle.)

4. Tape the rods or pencils to either side of the batteries. Snap the battery connectors onto the batteries and clip one alligator clip to each graphite rod.

A. Electrolysis of Water. Use your electrolysis unit to split water (H_2O) into hydrogen (H_2) and oxygen (O_2).

1. Obtain a petri dish. Fill it about half full with distilled or deionized water. Add 2 drops of phenolphthalein indicator and swirl gently to mix. This indicator is colorless in neutral or acidic solutions (excess H^+) and pink in basic solutions (excess OH^-). (Your instructor may provide a different indicator and, if so, will tell you the color changes.)

2. Immerse the tips of the graphite rods and observe carefully.   Do you see any evidence of a chemical reaction?

3. Add about one-half teaspoon (about 1 gram) of solid potassium nitrate, KNO_3. Jiggle the dish gently to aid in dissolving the KNO_3 crystals. The job of the KNO_3 is to be the electrolyte, i.e., to help the solution conduct electricity by providing K^+ and NO_3^- ions, <u>without</u> directly participating in any reactions.

4. Again immerse the graphite rods in the solution. Observe carefully for a couple of minutes. You should see evidence of chemical change. (Set the dish on a light-colored background to make the observations clearer.) Record what is happening at the electrode attached to the red wires and what is happening at the electrode attached to the black wires. Which one produces more gas bubbles? Which causes the indicator to change color?

B. Electrolysis of a Water Solution of Potassium Iodide, KI(aq)

1. Rinse the solution from the previous part down the drain with excess water. Then add about one-half teaspoon (about 1 gram) of solid potassium iodide, KI, to the dish.

2. Fill your petri dish at least half full with tap water. Jiggle the dish gently to aid in dissolving the KI crystals. The KI does two jobs here. It serves as the electrolyte, helping the water conduct electricity, but it also will be a reactant.

3. Immerse the graphite electrodes in the solution. Record your observations, noting what is happening at or near each of the two electrodes.

4. While the electrolysis is running, add 2 drops of phenolphthalein indicator directly next to each electrode. As noted previous, this indicator is colorless in neutral or acidic solutions (excess H^+) and pink in basic solutions (excess OH^-). Record any color changes you see and note whether this was at the red or black wire electrode.

Clean Up

1. Dump the petri dish contents in the designated waste container.
2. Return the electrolysis unit or disassemble it if instructed to do so.

Questions To Be Answered After Completing This Experiment

Write out answers to the following questions on a separate sheet of paper and hand it in along with the data sheet.

NOTE: The following questions are designed to aid in the interpretation of the experimental data and are an important part of this experiment.

1. Sketch a diagram of your Cu/Zn cell showing the metals and solutions involved as well as the voltmeter. Show the direction of electron flow out of one metal and through the voltmeter to the other metal. (Recall that the red-wire probe was on the metal consuming electrons.) Show the rest of the path that negative charge must follow to complete the circuit. Explain why the KNO_3 solution was needed.

2. In Part I, you measured the voltages of three voltaic cells. Rank these in order by the voltages they generated. Include in your rankings a standard Zn/MnO_2 dry cell (1.5 volts), similar to what you use in flashlights, etc.

3. Voltages can be increased by hooking together several cells in a series. How many of the Cu/Zn cells would you need to combine to get the same 1.5 volts you get from a purchased dry cell?

4. What voltage did your citrus cell give? How many of these would you need to get a voltage equivalent to a dry cell?

5. Imagine that you are marooned on an isolated island with a few other people. (You may remember a recent popular TV series.) After doing this experiment, what could you offer to the group that would be useful for supplying energy? Suggest something specific you could do with a few scraps of metal and food scraps or some nearby salt deposits.

6. In Part II-A, what two gases were the final products from the electrolysis of water? Write a balanced equation for the electrolysis of water. (Water is the only reactant, and the gases are products.)

7. The two half-reactions for the electrolysis of water are

$$4\,H_2O(l) + 4\,e^- \rightarrow 2\,H_2(g) + 4\,OH^-(aq) \quad \text{(2 moles of gas and a basic solution formed)}$$
and
$$2\,H_2O(l) \rightarrow O_2(g) + 4\,e^- + 4\,H^+(aq) \quad \text{(1 mole of gas and an acidic solution formed)}$$

Based upon your observations, which of the two half-reactions appears to be occurring at the black-wire lead? Explain your reasoning. Which gas formed at which color electrode? Explain how you decided.

8. In light of your answer to #7 and the equations for the half-reactions, which color electrode was supplying electrons from the batteries? What was the other electrode doing? Explain briefly.

9. In Part II-B, during the electrolysis of the KI solution, either H_2O or KI could react at each of the electrodes. Thus, although only one reaction will actually occur at each electrode, we must consider two possible reactions for the red electrode and two possible reactions for the black electrode. The half-reactions that could occur are the following:

at the electrode where electrons are consumed:

Either: $4\,H_2O(l) + 4\,e^- \rightarrow 2\,H_2(g) + 4\,OH^-(aq)$ (gas and basic solution formed)

Or: $K^+(aq) + 1\,e^- \rightarrow K(s)$ (metallic potassium formed)

at the electrode where electrodes are released:

Either: $2\,H_2O(l) \rightarrow O_2(g) + 4\,e^- + 4\,H^+(aq)$ (gas and acidic solution formed)

Or: $2\,I^-(aq) \rightarrow I_2(aq) + 2\,e^-$ (yellow-brown elemental I_2 formed)

   a. Based upon your observations, what formed at the black-wire lead? Explain your reasoning.
   b. Based upon your observations, what formed at the red-wire lead? Explain your reasoning.
   c. Now compare your electrolysis of water to that of the KI solution. Which product from the water electrolysis was also produced in the KI electrolysis?
   d. Which product from the water electrolysis was NOT produced in the KI electrolysis? What was formed instead? Electrolysis will always produce the products that are easiest to make from the starting materials. What can you conclude about how easy it is to remove electrons from iodine in KI versus how easy it is to remove electrons from oxygen in H_2O?

Name _____   Date _____

Partner _____   Lab Section _____

Data Sheet—Experiment 22

Part I: Electrochemical (voltaic) Cells

Zn/Cu cell voltage = _____   Red probe on which metal? _____

Zn/Ag cell voltage = _____   Red probe on which metal? _____

Cu/Ag cell voltage = _____   Red probe on which metal? _____

Citrus cell voltage = _____   Metals used _____

Part II: Electrolytic Cells

A. Electrolysis of water:

Observations at the black-wire electrode

Observations at the red-wire electrode

B. Electrolysis of potassium iodide solution:

Observations at the black-wire electrode

Observations at the red-wire electrode

Polymer Synthesis and Properties

Introduction

As explained in Chapter 9 of *Chemistry in Context,* fourth edition, nearly all of the thousands of known polymers are composed largely of carbon. Although these polymers have very similar elemental compositions, their properties vary widely. We can make polymers that are opaque or transparent, rigid or flexible, weak or strong, sticky or smooth. In this experiment, you will examine the properties of several polymers, including some that you will synthesize, and then connect polymer properties to their structures.

Background Information

The properties of a polymer depend not only upon the nature of its monomer, but also on the arrangement of the polymer strands relative to each other. In most polymers, the molecules are arranged randomly, as in *Figure 23.1*. Such polymers will be equally strong in all directions. In some polymers, however, the molecule strands are arranged parallel to each other, as in *Figure 23.2*.

Figure 23.1 Random polymer strands   **Figure 23.2** Parallel polymer strands

Changing from random to parallel strands can change the polymer's transparency, melting point, or other properties. We also expect the strength of a polymer with parallel strands to vary with direction. Pulling lengthwise is resisted by covalent bonds within a molecule strand. Pulling crosswise, however, simply moves one strand further away from another.

In some polymers, one molecule strand is only weakly attracted to the next strand. This results in relatively weak polymers. Stronger attractions between strands result in solid polymers that are stronger or more rigid and liquid polymers that are more viscous (syrupy).

In Part II, you will prepare nylon, a very strong fiber discussed in Section 9.6 of *Chemistry in Context*. You will do a condensation reaction of a di-acid chloride with a di-amine.

$$Cl-\overset{\overset{O}{\|}}{C}-(CH_2)_4-\overset{\overset{O}{\|}}{C}-Cl \ + \ H-\underset{\underset{H}{|}}{N}-(CH_2)_6-\underset{\underset{H}{|}}{N}-H \ \longrightarrow \ \left[\overset{\overset{O}{\|}}{C}-(CH_2)_4-\overset{\overset{O}{\|}}{C}-\underset{\underset{H}{|}}{N}-(CH_2)_6-\underset{\underset{H}{|}}{N}\right]_n \ + \ HCl$$

In Part III, you will prepare polyurethane foam. This polymer has multiple growth points on each strand, and thus, the polymer is able to branch and forms cross-linked strands. A gas produced by a side reaction causes the polyurethane to foam.

In Parts IV and V, you will add sodium borate to polyvinyl alcohol, creating cross-links, or bonds, between adjacent polymer strands, as shown in the equation below. More abundant and stronger cross-link bonds give polymers that are more rigid or more viscous.

Throughout this experiment, you will be asked to observe and record properties of polymers. The properties that are noteworthy will vary, but examples you should consider include

- color
- physical state (liquid or solid?)
- viscosity (how syrupy?)
- brittleness (would it break if hit hard?)
- elasticity (will it stretch and snap back?)
- adhesion (sticky or smooth?)
- tensile strength (how hard can you pull before it breaks?)

Overview of the Experiment

1. Observe properties of Teflon® tape and HDPE before and after stretching.
2. Prepare and observe properties of nylon.
3. Prepare and observe properties of polyurethane foam.
4. Prepare and observe properties of polyvinyl alcohol gel.
5. Measure effect of different amounts of sodium borate on polyvinyl alcohol gel viscosity.

Materials Needed

Chemicals	Equipment
• Teflon® tape (5 cm)	• 2 50-mL beakers
• HDPE film (5 cm x 10 cm)	• 1 100-mL beaker
• 10 mL nylon solution A (adipyl chloride in hexane)	• tweezers
• 10 mL nylon solution B (aqueous hexamethylenediamine)	• test tube
• 4 mL polyurethane A solution	• 3-oz paper cup
• 4 mL polyurethane B solution	• wooden splint or applicator stick
• food coloring, optional	• metal spatula
• 100 mL 4% polyvinyl alcohol solution	• ruler
• 10 mL 4% sodium borate solution	• marker
	• steel ball (1/2" ball bearing)
	• stopwatch or watch with second hand

SAFETY NOTES
- Avoid getting the solutions used to prepare nylon or polyurethane on your skin. If you do get these on your skin, wash them off with large amounts of water.

- The polymers produced may be handled directly, but you should wash your hands afterward.

EXPERIMENTAL PROCEDURE

I. Examination of Polymer Films

1. **Teflon**. Obtain a piece of Teflon® tape (roughly 5 cm long). Observe the properties of this polymer. (See the Background section for suggested properties to consider.) Test the strength of the polymer by gently pulling first along the length of the polymer strip and then side-to-side. Record your observations.

2. **HDPE**. Obtain a piece of high-density polyethylene film (roughly 5 cm x 10 cm) that has been cut from a grocery-store bag.

   a. Observe and record the properties of this polymer. Test the strength of the polymer by gently pulling first lengthwise and then side-to-side. Record your observations.

   b. Now grasp the film by its shorter sides. Pull firmly and slowly until the middle of the strip has stretched significantly. Observe and record the properties of this stretched section. Compare the width of the stretched section to that of the unstretched end portions. Test the strength of the stretched portion both lengthwise and crosswise. Record your observations.

II. Synthesis of Nylon

1. Obtain 10 mL of nylon solution A (adipyl chloride in hexane) in a 50-mL beaker. In a second 50-mL beaker, obtain 10 mL of nylon solution B (hexamethylenediamine in water).

2. Tip the beaker containing solution B at a slight angle and then *slowly* add solution A by gently pouring it down the side of the beaker. Solution A should form a separate layer on top of solution B. Do NOT stir or mix. A film of nylon will form between the layers.

> **SAFETY NOTE**: Do not touch the nylon with your hands until it has been thoroughly rinsed with water. Hydrochloric acid is formed as a reaction by-product.

3. Use tweezers to grasp the film. Gently pull it up and out of the beaker. Wrap the film around a test tube and then rotate the test tube to gradually pull the rest of the film from the beaker. If you are careful, you can get one long continuous strand. If the strand breaks, simply grab it again with tweezers and wrap it around the test tube. Continue pulling nylon film until the solutions are used up.

4. Rinse your nylon thoroughly under a gentle stream of tap water. Slide the nylon off of the test tube. Observe and record the properties of the wet nylon.

5. Spread some of your nylon strands out on a paper towel to dry. Continue with other parts of the experiment but return to observe and record the properties of the nylon after it dries.

6. When done, solid nylon strands may be placed in the trash. Dispose of excess solutions as directed by your instructor.

III. Synthesis of Polyurethane Foam

1. Obtain 4 mL of polyurethane solution A in a 3-oz. paper cup. Be patient as you pour so that you get as much of the viscous liquid into the cup as possible. If desired, add 2 drops of food coloring. Mix with a wooden splint or disposable applicator stick.

2. Add 4 mL of polyurethane solution B to the paper cup. Use the wooden splint to <u>thoroughly</u> mix the liquids. Once the liquids are well mixed, remove the mixing stick and allow the cup to sit undisturbed.

3. The reaction may take a minute or two to begin. Once it does, record your observations. You may gently touch the outside of the cup as the reaction proceeds but do not touch the polymer itself yet.

4. After about 5 minutes, the reaction will be complete, and you may touch the polymer, which should no longer be sticky. Observe and record the properties of the polyurethane foam.

5. When done, the polyurethane foam may be placed in the trash.

IV. Synthesis of Polyvinyl Alcohol Gel

1. Obtain 50 mL of 4% polyvinyl alcohol solution in a 100-mL beaker. (Pure polyvinyl alcohol is a white solid, but it has been dissolved in water here.) Observe and record the properties of this polymer solution.

2. Obtain 5 mL of 4% sodium borate solution. Add this to the polyvinyl alcohol with stirring. Continue stirring until the mixture becomes homogeneous. This may take several minutes. You may wish to use a metal spatula rather than a glass rod to stir as the mixture becomes very thick.

3. Remove the polymer from its beaker and examine it. You can touch the polymer with your hands, just be sure to wash your hands afterwards. Try forming a ball with the polymer. Does it bounce? Does it remain round? Does it stretch? Observe and record the properties of the polyvinyl alcohol gel.

4. When done, the polyvinyl alcohol gel may be placed in the trash.

V. Viscosity of Polyvinyl Alcohol Gels

In this portion of the experiment, you will determine how the amount of sodium borate solution added to polyvinyl alcohol affects the viscosity of the polymer. Your instructor will assign you a specific amount of sodium borate to add. Work carefully since your results will be combined with those of the rest of the class.

1. Obtain 50 mL of 4% polyvinyl alcohol solution in a 100-mL beaker as before.

2. Obtain your assigned volume of 4% sodium borate solution (from 0 to 5 mL). Add enough water to bring this solution to 5 mL. Mix.

3. Note the time and then add the sodium borate solution to the polyvinyl alcohol solution. Mix thoroughly. It is important to get this gel as homogeneous as possible. A metal spatula may be useful to break up solid lumps. Periodically mix and break up lumps for 5 minutes.

4. After mixing for 5 minutes, transfer your gel into a 50-mL beaker. Use a marker and ruler to place two lines 30 mm apart on the side of the beaker. The top line should be somewhat below the surface of the gel and the bottom line should be somewhat above the bottom of the beaker.

5. Hold a steel ball above your gel so that it just touches the surface. Release the ball and record the time in seconds that it takes the ball to move from the top line to the bottom line.

6. As directed by your instructor, post your data (volume of sodium borate and time for the ball to drop) for use by the rest of the class.

7. Dispose of your gel mixture as directed by your instructor.

8. **Graph your data**. Use the graph paper provided to make a graph of the class data, recording time for the steel ball to drop (in seconds) on the *y*-axis versus volume of sodium borate added on the *x*-axis. Draw a smooth curve through your data points.

Questions To Be Answered After Completing This Experiment

Write out answers to the following questions on a separate sheet of paper and hand it in along with the data sheet.

---

1. In Teflon® tape, are the polymer stands arranged randomly or in parallel? Describe your evidence.

2. When high-density polyethylene films are stretched, they form "necks." Describe how the properties you observed changed in the necked region. What happened to the arrangement of the polymer strands?

3. Nylon is often used as fibers (threads). What lengthwise and crosswise strength do you think would be important for a fiber? Explain your reasoning. Did your observation of nylon's properties match this expectation?

4. Polyurethane solutions A and B are sometimes injected directly into the walls of older homes and allowed to foam in place. Why?

5. Both polyurethane foam and Polyvinyl alcohol gel contain cross-links. What are cross-links? How do the strengths of cross-links in these two polymers compare to each other? Describe your evidence.

6. Describe the relationship between the viscosity of polyvinyl alcohol gel and the amount of sodium borate added. Explain the connection to cross-linking.

Name _____   Date _____

Partner _____   Lab Section _____

Data Sheet—Experiment 23

Part I: Examination of Polymer Films

1. Properties of Teflon® tape:

2a. Initial properties of high-density polyethylene film:

2b. Properties of stretched polyethylene film:

Part II: Synthesis of Nylon

1. Properties of wet nylon strand:

2. Properties of dried nylon strand:

Part III: Synthesis of Polyurethane Foam

1. Observations as the reaction proceeds:

2. Properties of the final polyurethane foam:

Data Sheet—Experiment 23, page 2

Part IV: Synthesis of Polyvinyl Alcohol Gel

1. Properties of original polyvinyl alcohol solution:

2. Properties of the final gel:

Part V: Viscosity of Polyvinyl Alcohol Gels

      Assigned volume of sodium borate _____

      Drop time for steel ball (seconds)   _____

      Properties of the gel:

Classification and Identification of Common Plastics

Introduction

Of the more than 60,000 plastics manufactured, six of them make up 70% of the plastics used in this country. These six are low-density polyethylene, high-density polyethylene, polyethylene terephthalate, polyvinyl chloride, polystyrene, and polypropylene. They are encountered by most people every day in familiar products, yet most people are unaware of their specific uses and properties. In fact, most people would be unable to identify them in any specific way. In this experiment, you will examine some properties of these six common plastics and use a few simple tests to develop a classification and identification scheme.

Background Information

The plastics industry has adopted a code for packaging materials that can be used to identify the type of plastic in a plastic container. The idea behind the symbol is to make recycling easier by making the identification of the plastics easier. The symbol on the bottom of many containers is a triangle of arrows chasing each other with a number in the middle of the triangle.

The code for these symbols is as follows:

No.	Name	Abbrev.
1	Polyethylene terephthalate	(PET)
2	High-density polyethylene	(HDPE)
3	Polyvinyl chloride	(PVC)
4	Low-density polyethylene	(LDPE)
5	Polypropylene	(PP)
6	Polystyrene	(PS)
7	Other (includes composites and mixtures)	

Compliance in labeling is voluntary, and not all plastics have an identification code symbol. Without code numbers these plastics are difficult to separate or classify by appearance. However, there are some general rules: (1) Opaque translucent plastics (such as milk containers) are most likely high-density polyethylene (HDPE); (2) containers used for carbonated drinks are usually polyethylene terephthalate (PET).

In this experiment, you will use four simple tests to aid in classification and identification of the six types of plastics. The first test will be to determine the relative densities of the plastics by checking to see whether the samples float or sink in three liquids of differing densities.

The second test will be to melt the plastic. All six of these common plastics melt reversibly, which means that when they are cooled, they harden and may regain their original properties. If a plastic sample does not melt, it is a thermosetting plastic. Thermosetting plastics, such as melamine (used in high-quality plastic dinnerware) or the plastic used in electrical components, do not melt cleanly and reversibly, but tend to char instead.

The third test, burning the plastic, must be performed only in a fume hood. All common plastics burn (some only if held directly in the flame), but with slightly different characteristics and different noxious fumes. The vapors given off from the burning plastic may have different properties depending on the plastic.

The last test, the copper-wire test, will be used to determine if a halogen, such as chlorine, is part of the polymer.

Overview of the Experiment

1. Obtain known samples of the 6 plastics.
2. Determine the relative densities of the samples.
3. Test each sample for melting.
4. Test each sample to see if it burns.
5. Test each sample with the copper-wire test.
6. Devise a classification scheme and a way to identify each plastic.
7. Test your scheme with some known samples provided by the instructor.
8. Test your scheme with some unknown samples that you brought from home.

Materials Needed

Chemicals	Equipment
• known samples of all six types of plastic • plastic samples of unknown composition provided by the instructor • plastic samples of unknown composition that you bring from home • 1:1 mixture of 95% ethanol and water • 10% NaCl solution in water	• 3 test tubes with a rack or stand • microburner or Bunsen burner • matches • 6-inch piece of copper wire and a cork • metal spatula or razor blade • few strips of blue litmus paper • pair of forceps, tongs, or pliers • scissors • glass stirring rod

---

SAFETY NOTES

• Exercise caution when using an open flame. Long hair should be tied back, and extremely loose sleeves should be avoided.

• Pieces of hot molten plastic would cause severe burns if they were to drop onto your skin or clothing.

EXPERIMENTAL PROCEDURE

I. Density Tests

Three liquids of differing densities will be used, as shown in the following chart.

Liquid	Density g/cm^3
1:1 95% Ethanol/ water	0.94
Water	1.0
10% NaCl in water	1.08

1. Obtain 3 test tubes. Pour about 5 ml of the 1:1 95% ethanol/water mixture in the first test tube and label it with its density (from the chart above). In the same fashion, put water in the second tube and 10% NaCl solution in the third tube. Be sure each tube is labeled.

2. Obtain two narrow strips of each of the 6 types of plastic. Cut *one* strip of each plastic into 3 small pieces.

3. For one of the plastics, place a piece into each of the three test tubes containing the density test liquids. Push each piece under the liquid surface with a glass stirring rod or the end of a pencil. If the sample floats, it has a density lower than that of the liquid. If it sinks, it has a density greater than that of the liquid. Record your observations on the data sheet.

4. In the same fashion, test each of the other plastics and record your observations.

5. Finally, use your observations to sort the six plastics by their densities, filling in the table and the ranking list on the data sheet.

II. Melt Test

1. Place a small sample of each plastic, one at a time, on the end of a metal spatula and hold the end of the spatula over a light blue microburner flame.

2. Heat slowly ("cook the plastic") and observe the plastic as it warms and finally melts. **Do not** heat strongly enough for the plastic to catch fire. Enter your observations in the data table.

3. Cool the sample and examine it for appearance and flexibility by bending it. Enter your observations in the data table.

4. The melted plastic can still be used for the tests that follow.

III. Ignition Test   CAUTION: Do this only in a fume hood. Toxic fumes are produced.

1. Place a microburner or Bunsen burner in a fume hood. Light it and adjust to a small flame.

2. Place a large beaker of water in the hood.

3. Hold one end of a small strip of plastic in a pair of tongs, forceps, or pliers and place it directly in the flame. Observe the color of the flame and its characteristics. (Is a lot of smoke or visible vapor given off? Does the plastic continue to burn after it is removed from the flame?) Record your observations in the data table.

4. Test the vapors given off for acidic properties by holding a piece of <u>wet</u> litmus paper in the vapors from the burning plastic. If the paper turns red, acidic fumes are being formed as the plastic burns. Record your observations in the data table.

5. Extinguish the burning plastic by dropping it into the beaker of water.

6. Repeat steps 3–7 for each of the other plastics.

IV. Copper-Wire Test CAUTION: Do this only in a fume hood. Toxic fumes are produced.

1. Push the end of a 6-inch length of copper wire into a small cork.

2. Use the cork as a handle and heat the free end of the wire in a burner flame until the flame has no green color.

3. Touch the hot copper wire to the plastic you are testing and then return the wire end to the flame. A tiny bit of plastic should be picked up by the hot wire. Return the wire end to the flame. When the tip of the wire is put in the flame, watch for a slight flash of luminous flame. This indicates that you have correctly picked up a little bit of plastic on the wire.

4. Watch for the appearance of a green flame or green color in the flame when the plastic is heated in the flame. The green color indicates the presence of chlorine in the plastic.

V. A Puzzle for You to Solve

Devise a scheme for identifying the six plastics, based on the simple tests you have performed. The challenge is to find the minimum number of tests that will correctly identify the plastics if you are given any one of the plastics as an unknown. After you have devised your scheme, use it to identify at least two plastic samples that you have brought from home.

Finally, two "unknown" plastic samples will be given to you to identify. Use your scheme to determine their identities.

> ### Questions To Be Answered After Completing This Experiment
> Write out answers to the following questions on a separate sheet of paper and hand it in along with the data sheet.

1. Outline your scheme for identifying the six common plastics.

2. Suppose you had to add two other plastics to your scheme, polymethymethacrylate (density 1.18–1.20 g/cm^3) and poly-4-methyl-1-pentene (density 0.83 g/cm^3). Where would they fit in your scheme?

3. Polyethylene terephthalate (PET) is the most valuable waste plastic at the present time. Suggest a way to separate it commercially from other waste plastics.

4. Why are plastic recyclers very concerned about identifying the different polymers and not mixing them together?

5. Since waste plastic consists mostly of hydrocarbon compounds, it has been suggested that waste plastic could be used as fuel. Based on your observations in this experiment, do you think this is a reasonable suggestion? Would some be more dangerous to burn than others? Defend your answer.

Name _____   Date _____

Partner _____   Lab Section _____

Data Sheet—Experiment 24

Unknown 1 number_____

Unknown 2 number_____

I. Density Tests (write in either "sinks," "floats," or "can't tell")

Type of plastic	1:1 Ethanol/H_2O density 0.94 g/cm^3	Water density 1.0 g/cm^3	10% NaCl solution density 1.08 g/cm^3
PET			
HDPE			
PVC			
LDPE			
PP			
PS			
Sample from home #1			
Sample from home #2			
Unknown 1			
Unknown 2			

Plastic Density Categories: (place each plastic into one of these categories)

Less than 0.94 g/cm^3	Less than 1.0 g/cm^3	Less than 1.08 g/cm^3	More than 1.08g/cm^3

Ranking of densities:

   (lowest) _____ _____ _____ _____ _____ _____ (highest)

Name _____   Date _____

Partner _____   Lab Section _____

Data Sheet—Experiment 24, page 2

Melting and Ignition Tests

Plastic	Melting test	Ignition test	Copper-wire test
PET		Acidic smoke? yes/no _____	
HDPE		Acidic smoke? yes/no _____	
PVC		Acidic smoke? yes/no _____	
LDPE		Acidic smoke? yes/no _____	
PP		Acidic smoke? yes/no _____	
PS		Acidic smoke? yes/no _____	
Sample from home		Acidic smoke? yes/no _____	
Sample from home		Acidic smoke? yes/no _____	
Unknown 1		Acidic smoke? yes/no _____	
Unknown 2		Acidic smoke? yes/no _____	

Identification of Unknown Samples

Sample from home #1: Description _____ Identity _____

Sample from home #2: Description _____ Identity _____

Unknown 1 number _____   Identity _____

Unknown 2 number _____   Identity _____

Identification of Analgesic Drugs
by Thin-Layer Chromatography

Introduction

In this experiment, you will utilize thin-layer chromatography to identify the analgesic compound(s) present in an over-the-counter (OTC) painkiller preparation. OTC analgesic preparations are widely sold and used in the United States. Although extremely useful and effective, they are also powerful drugs that can be very harmful if misused. Because such analgesics are so commonly available in homes, there is a danger of accidental overdose by young children who discover the innocuous-looking white pills. In cases of accidental poisoning, there may be need for quick identification of what is in an unknown tablet. This experiment demonstrates one such method.

Analgesic compounds are discussed in Sections 10.3 and 10.4 of *Chemistry in Context,* fourth edition. Aspirin (acetylsalicylic acid) is by far the most common. Other common non-prescription analgesics include acetaminophen (Tylenol) and ibuprofen. Caffeine is sometimes added to analgesic tablets to overcome drowsiness. Tablets also contain binders such as starch and sugars. "Buffered" aspirin contains a base such as magnesium hydroxide or calcium carbonate to neutralize the acid.

Background Information

Chromatography is a technique used to separate and identify individual components in a mixture. Its name comes from the Greek word for color because the earliest applications of chromatography dealt with mixtures of colored substances such as plant pigments. A variety of chromatographic techniques have been developed, of which gas chromatography (GC) is probably the most widely used. (Gas chromatography is described briefly in the text in Section 5.16.3.) All chromatographic techniques make use of the fact that components of a mixture (either gaseous or in solution) tend to move at different speeds along a coated tube or flat surface. The different rates of movement are the result of differing attractions of the components to the coating material compared to the tendency of the components to remain in the moving fluid (either gas or liquid).

Thin-layer chromatography (TLC) is one of the easiest chromatographic techniques. A thin layer of a suitable solid substance is coated on a sheet of glass or plastic. By immersing one edge of the sheet in an appropriate liquid solvent, the solvent is drawn up the sheet by capillary action, and the compounds of interest are carried along at differing rates. This is commonly called "developing" the plate.

Unless the compounds of interest are colored, some method is necessary to make them visible. One method is to expose the developed plate to some compound that will react chemically with the spots to make them visible. Another method is to examine the plate under ultraviolet light to see if any of the components are fluorescent. An ingenious alternative to that is to incorporate a fluorescent dye into the plate so that when the plate is examined under a UV lamp, the plate will glow (fluoresce) everywhere *except* where the component spots are.

Overview of the Experiment

1. Prepare a developing container.
2. Spot known and unknown samples on a chromatographic sheet.
3. Develop the chromatogram.
4. Calculate R_f values.
5. Identify components in unknown sample.

Materials Needed

Chemicals	Equipment

• solutions of analgesics in methanol
    • aspirin
    • acetaminophen
    • ibuprofen
    • caffeine
    • (optional) mixture of all 4
• unknown OTC analgesic tablet
• methanol
• solvent mixture
	• large beaker, 400–600 mL
	• plastic sandwich wrap to cover beaker
	  (or aluminum foil or screw-cap bottle)
	• glass capillary tubes for spotting
	• plastic TLC sheet, about 5 cm x 10 cm
	• small ruler (metric scale) and pencil
	• mortar and pestle
	• ultraviolet lamp (shared by whole class)

EXPERIMENTAL PROCEDURE

1. Obtain a large beaker (400–600 mL) and a piece of plastic wrap or aluminum foil to cover it. Alternatively, a wide-mouthed jar with screw-cap lid may be used. The beaker or jar should be large enough so that the chromatographic sheet can lean against one side (see *Figure 25.2*).

2. Add enough of the solvent mixture (containing 25 parts ethyl acetate, 1 part ethanol, and 1 part acetic acid) to give a thin layer of solvent in the bottom of the container. To provide an atmosphere saturated with solvent inside the container, place a piece of filter paper around the inside surface of the container, extending into the solvent (see *Figure 25.2*). Then cover the container with the plastic wrap, foil, or screw cap and set it aside while preparing the chromatographic sheet (steps 3–6).

3. Obtain an unknown analgesic tablet (or a portion of one). Using a mortar and pestle, crush a portion of a tablet to a fine powder. (Alternatively, this can be done in a small test tube with a glass rod, but it must be done carefully to avoid breaking the test tube.) Add a few milliliters of methanol to the powder and stir; then allow the mixture to settle for a few minutes.

4. Obtain a piece of plastic-backed chromatographic sheet. Unless you are told otherwise, this is a special kind of TLC sheet that contains a fluorescent dye to help make the results more visible. The white coating may flake off if the sheet is not handled carefully. Handle it only on the edges. The backing is flexible, but the sheet should not be bent excessively. Using a pencil (not a pen), draw a *very light* line across the sheet (Short dimension) about 1 cm from one end (*Figure 25.1*). Then make 5 small light marks at 1-cm intervals along the line. These are the

points at which the samples will be spotted. (Your instructor may suggest different spacing, depending on the size of TLC sheet and the number of spots used.) Label each point with a penciled number or abbreviation and record this code on the data sheet.

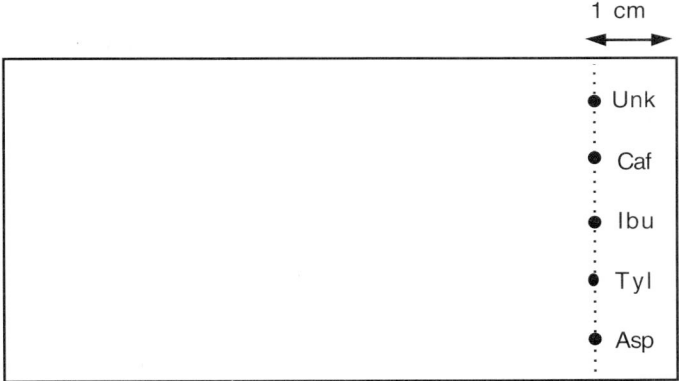

Figure 25.1 Layout of TLC sheet

5. Samples will be spotted onto the chromatographic sheet, using glass capillary tubes drawn down to a very small opening. (The tubes are quite fragile.) Before spotting the sheet, you should practice putting *very small* spots of solution on a piece of scrap paper. To do this, dip the tip of a capillary into a solution, then very gently touch the tip to the paper for a brief moment. The spots should be as small as possible in order to minimize tailing and overlapping when the chromatographic sheet is "developed." (If a more intense spot is desired, it is preferable to let the spot dry and re-spot in the same location.)

6. Solutions of the 3 analgesic compounds plus caffeine will be available in the lab with a glass capillary tube in each. Carefully place small spots of the four solutions at four pencil marks. (**Optional:** your instructor may ask you to add another spot with a mixture of all four compounds.) Finally, spot a sample of the clear solution from your unknown tablet onto the chromatographic plate, using a separate glass capillary tube. Allow the solvent to evaporate. A hair dryer or oven can be used to speed up the drying process. **Optional:** If two students are working together, you may wish to prepare a second TLC sheet so that each of you will have one to keep. The second one can be developed (steps 7 and 8) in a separate beaker or after the first one is done.

7. When the spots are dry, place the sheet in the developing container (*Figure 25.2*). Check to be sure that the bottom edge (near the spots) is in the solvent but that the spots are above the solvent. Also be sure that the filter paper does not touch the chromatographic sheet. Then cover with plastic wrap, foil, or a screw-cap lid and watch carefully. The liquid will slowly move up the TLC sheet by capillary action.

8. When the front edge of the liquid has moved about 80–90% of the way to the top of the sheet, remove the sheet. *Immediately*, while the sheet is still wet, draw a pencil line on the sheet to show the top edge of the liquid. Then lay the sheet on a clean surface in a fume hood or well-ventilated area and allow the solvent to evaporate until the sheet appears dry.

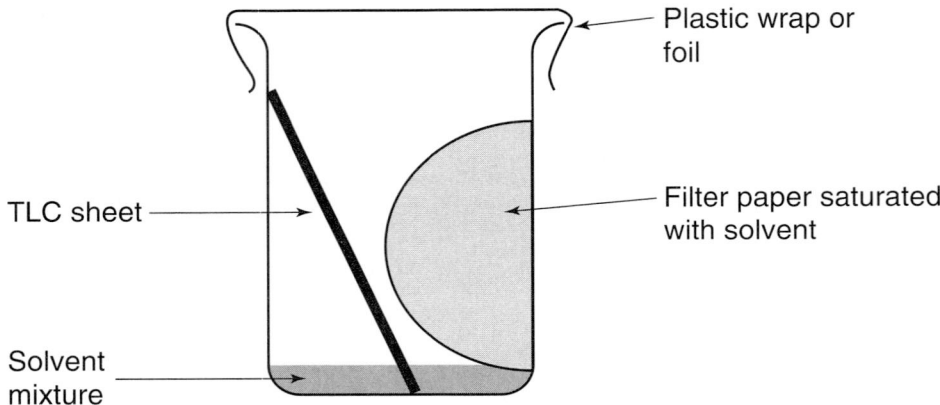

Figure 25.2 The assembled TLC developing chamber

9. The spots are unlikely to be visible to the naked eye, but they should be quite visible when viewed under an ultraviolet (UV) lamp. It may be necessary to do this in a location with subdued light. While observing under the UV lamp, draw a light pencil line around each spot.

> **Caution**: UV radiation is harmful to your eyes. Do not look directly at the UV lamp.

10. It is quite likely that you can see the crucial information at a glance: (a) Did the three pure analgesic compounds and caffeine each produce a single spot, and did the spots move different distances? (b) If you used a mixture of analgesics, did it produce the expected number of spots, and do their locations match those of the pure components? (c) Finally, and of most interest, how many spots did the unknown analgesic produce? Can you identify the compound(s) present in the unknown tablet?

11. A more elegant way of interpreting the results is to calculate a *retention factor* (R_f) for each spot. The retention factor is the ratio of the distances traveled by the component and the solvent.

$$R_f = \frac{\text{distance traveled by the substance}}{\text{distanced traveled by the solvent}} = \frac{\text{distance to center of spot}}{\text{distance to solvent front}}$$

a. Obtain a short ruler marked in centimeters and millimeters. For each spot, carefully measure the distance (*in millimeters*) from the starting line to the solvent front and also from the starting line to the middle of the spot. Some judgment may be necessary in choosing the middle of the spot. Record these numbers on the data sheet. If more than one spot is present for the unknown, make these measurements for each spot.

b. Calculate the R_f value for each spot, dividing the distance to the center of the spot by the distance to the solvent front. This number should be less than 1, since the spot did not move as far as the solvent. Record this ratio on the data sheet, writing it as a decimal fraction rounded to two digits after the decimal point.

Questions To Be Answered After Completing This Experiment

Write out answers to the following questions on a separate sheet of paper and hand it in along with the data sheet.

---

1. Why do you think it is necessary to maintain a saturated vapor atmosphere in the container used for development? What do you predict would happen if this were not done?

2. Suggest possible advantages and disadvantages of using a longer (taller) TLC sheet.

3. Why do you think it was important to use a very small amount of sample when spotting the plate?

4. The relative movement of components is controlled partially by the polarity of the molecules. The TLC sheet is coated with a highly polar substance, whereas the solvent mixture has a much lower polarity. From your chromatographic results, predict the relative polarities of aspirin, acetaminophen, ibuprofen, and caffeine. Explain your reasoning.

5. Do you expect that changing the solvent will change the R_f value for a given component? Explain your reasoning.

6. If two components have an identical R_f value, does this mean they necessarily have the same structure? Explain why or why not.

7. In an effort to identify an unknown, some students obtain the TLC plate shown at the right.

   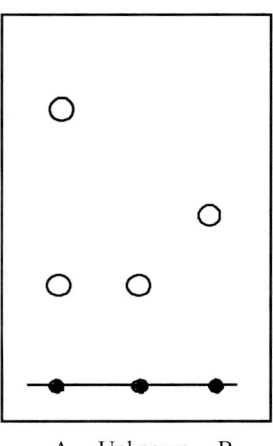

   a. How many compounds were in Sample A? In Sample B? In the unknown?

   b. A student concludes that the unknown is the same as Sample B because of the number of spots. Is this a valid conclusion? Explain.

   c. A second student concludes that the unknown is the same as Sample A because they both have spots with R_f values of 0.3. Is this a valid conclusion? Explain.

   d. Propose the most reasonable conclusion you can for the identity of the unknown. Explain your reasoning.

Notes

Name _____   Date _____

Partner _____   Lab Section _____

Data Sheet—Experiment 25

Sample	Label on TLC sheet	Distance to center of spot	Distance to solvent front	Calculated R_f
Aspirin				
Acetaminophen				
Ibuprofen				
Caffeine				
Mixture (optional)				
Unknown				

What compounds are present in the unknown tablet?

Tape the TLC sheet here. Cover the whole sheet with clear tape to protect it.

Synthesis of Aspirin

Introduction

This experiment is a representative example of a chemical synthesis; in this case, it is the synthesis of a common analgesic drug. Of all the drugs discussed in Chapter 11 of *Chemistry in Context*, only aspirin lends itself to a simple one-step synthesis. Although the other two common nonsteroid anti-inflammatory drugs, acetaminophen and ibuprofen, are prepared in tonnage quantities by commercial processes, their syntheses require more than one step and are beyond the scope of this course. Aspirin, on the other hand, can be easily made from salicylic acid by a reaction known as esterification. The synthesis method you will use is very similar to the original synthesis of aspirin carried out by Felix Hoffman in 1893.

Background Information

Hoffman's method for producing aspirin involved reacting salicylic acid with acetic acid in the presence of sulfuric acid as a catalyst.

Salicylic acid   Acetic acid   Aspirin (acetylsalicylic acid)

Instead of using acetic acid, you will use acetic anhydride, a compound that produces acetic acid when it reacts with water. Acetic anhydride is prepared by combining two acetic acid molecules.

Acetic acid   Acetic acid   Acetic anhydride

Acetic anhydride works better in the aspirin synthesis than acetic acid and requires less catalyst. The reaction is as follows:

Salicylic acid   Acetic anhydride   Aspirin (Acetylsalicylic acid)   Acetic acid

After the aspirin is synthesized, it must be isolated and purified. The impurities in the crude aspirin are sulfuric acid, unreacted acetic anhydride, acetic acid, and unreacted salicylic acid. Fortunately, the first three compounds are very soluble in water and easily washed away from the mixture during the isolation of the crude (unpurified) aspirin. However, salicylic acid and aspirin are only sparingly soluble in water, and any unreacted salicylic acid may be a contaminant in the crude product mixture. A technique known as **recrystallization** will be used to separate the aspirin from the unreacted salicylic acid. This technique involves dissolving the crude mixture in a suitable warm solvent. The solution is cooled to a point at which one of the components of the mixture crystallizes out of solution while the other component (the more soluble one) remains dissolved. The crystallized component can then be collected by filtering to remove it from the liquid.

Note: This experiment requires more equipment and more steps than most of the previous laboratory exercises. Therefore, it is important for you to read the instructions carefully and be sure you understand them before proceeding.

Overview of the Experiment

1. Collect the necessary equipment and chemicals.
2. Combine the salicylic acid, acetic anhydride, and sulfuric acid.
3. Heat the chemicals in a water bath.
4. Collect the crude product by vacuum filtration.
5. Purify the product by recrystallization.
6. Compare the properties of aspirin (crude and purified) with salicylic acid

Materials Needed

Chemicals	Equipment and glassware
• 2 grams of salicylic acid • 3 ml of acetic anhydride • 10 drops concentrated sulfuric acid • 6 mL of ethanol • crushed ice for an ice bath • glycerol for melting point bath	• 125-mL Erlenmeyer flask • 10-mL graduated cylinder • 50-ml Erlenmeyer flask • 250-ml beaker for a hot-water bath • small Büchner funnel or Hirsch funnel • filter paper to fit the funnel • filter flask, tubing, and filtering gasket • glass stirring rod • watch glass • beaker for ice bath • 3 melting-point capillary tubes • thermometer • large test tube • small Bunsen burner and matches

SAFETY NOTES: Both sulfuric acid and acetic anhydride are very corrosive acids. In addition, acetic anhydride has irritating fumes.

• Eye protection is absolutely essential.

• If you spill acid on yourself, wash it off with large amounts of water. Keep the portion of you that has the acid on it under running water for at least 5 minutes.

• If you spill acid on your work space, carefully wipe it up with a paper towel. Rinse the towel with water and then discard it in an appropriate container.

EXPERIMENTAL PROCEDURE

If laboratory time is limited, you may be instructed to omit Part II and/or Part III.

I. Synthesis of Aspirin

1. Half fill a 250-ml beaker with hot water and put it on a hot plate. This will serve as a hot-water bath.

2. Weigh 2.1 grams of salicylic acid (0.015 mole) into a <u>dry</u> 125-mL Erlenmeyer flask. (It is crucial that the flask be dry.   Any traces of water will interfere with the synthesis.) Record the appearance of the salicylic acid.

CAUTION: Do steps 3–8 in a fume hood.

3. In the fume hood, carefully measure 3 ml of acetic anhydride in a <u>dry</u> 10-ml graduated cylinder.

4. Add the acetic anhydride to the salicylic acid. Swirl to mix the liquid with the solid.

5. Carefully add 3 drops of sulfuric acid to the mixture.

6. Put the flask into the hot-water bath (which should be approximately boiling). Heat for 10 minutes. If necessary, stir the mixture with a glass rod to break up any lumps. Be sure you have heated for at least 5 minutes past the point where most of the solid has dissolved.

7. Remove your flask from the hot water. Add 2 mL of distilled water, a few drops at a time, with continuous swirling. *Do not add the water all at once.* The water is reacting with leftover excess acetic anhydride. Solid crystals of aspirin may start to form at this point.

8. Then add 10 grams of crushed ice to your flask and swirl. You should have a white solid product (aspirin) at this point. If not, consult your instructor before proceeding, especially if you see an oily liquid phase rather than a solid product. (Scratching the inside of the flask below the liquid level with a glass stirring rod will usually cause the oil to crystallize.   This may take several minutes. Alternatively, you can reheat the flask in a hot-water bath until all the oil disappears, then transfer the flask to an ice bath [next step] and swirl.)

9.   Return to your regular workbench. Cool your flask in an ice/water bath that you construct in a beaker. Use roughly a 2:1 mix of ice and water. (An optimal ice/water bath is mostly ice but with enough water to produce a slurry of ice and water.) Continue to cool the flask for 5–10 minutes with occasional swirling or stirring.

10.   While the reaction mixture is cooling, prepare a vacuum filtration setup such as the one shown in *Figure 26.1*. Your equipment may look slightly different from this. Your instructor will demonstrate how to use it. (It may be possible for several teams to share one setup.)

Figure 26.1 Diagram of the filtration setup

11.   Attach one end of the tubing to the side arm of the filter flask and the other end to the aspirator side arm on a water faucet.

12.   Attach a small Büchner funnel or Hirsch funnel to the filter flask and put a piece of filter paper in the funnel. The filter paper should lay flat and just cover all the holes in the bottom of the funnel.

13.   Wet the paper with a few drops of water.

14.   Turn the water full on. (Depending on the sink setup, you may need to add some sort of splash arrester in order to avoid being drenched.) When the water is on, there should be enough suction so that the filter paper is held firmly to the funnel.

15. Quickly pour the cooled slurry of product onto the center of the filter paper. If your setup is working correctly, the water should be sucked through the funnel, leaving a white pasty product on the filter paper.

16. Rinse the reaction flask with 2–3 ml of *ice-cold* water and pour the rinses through the funnel.

17. Keep the suction on for several minutes so that air is drawn through the solid until the product is nearly dry. Spreading the aspirin evenly on the filter paper will help speed the drying.

18. When the product is dry, detach the tubing from the side arm and then shut off the water. (Do NOT shut off the water first.)

19. Transfer the solid to a preweighed watch glass by inverting the funnel over the watch glass and tapping gently. Reweigh and record the mass. You should have about 2 grams of product.

II. Purifying the Aspirin by Recrystallization

At this point, the aspirin you obtained is somewhat wet and probably still contaminated with some salicylic acid. You will further purify it by a technique known as recrystallization.

> **CAUTION:** The next step uses ethanol, which is flammable. Be sure that there are no open flames anywhere in the lab.

1. Save a small amount of your crude aspirin for later comparison. Place the rest of it into a 50-mL Erlenmeyer flask. Add 6 mL ethanol. Heat this flask in a hot-water bath on a hot plate, preferably in a fume hood. The solid should all dissolve.

2. Add 10 mL distilled water to the flask and reheat in the hot-water bath until all of the solid again dissolves.

3. As soon as the solid has dissolved, take your flask back to your regular workstation and let it cool slowly to near room temperature. Do not cool rapidly with ice water. Crystals of product should form during the cooling. If not, consult your instructor.

4. Finally, cool the flask for 5 more minutes in an ice/water bath with occasional stirring. (**Warning:** Do not let the flask tip over in the ice/water bath. If it does, you will lose your product!)

5. Collect the product by suction filtration as described above.

6. Spread your product out to dry on a watch glass or paper towel.

At this point, your instructor will tell you whether or not to leave the product to dry thoroughly until the next laboratory period (and then completion of the experiment). If not, you can let it dry for a short time while you clean up your equipment and answer part of the questions at the end.

7. Carefully observe and describe this purified aspirin product. (This can be done while it is drying or afterward.) Does it appear different from the salicylic acid that you started with? If possible, examine a few crystals under a microscope or with a magnifying lens and observe the crystal shapes. Also examine the crude product. Record your observations in the data table.

8. Finally, determine the mass of purified aspirin by weighing it in a preweighed or "tared" vial or beaker. Many laboratory balances permit taring, which means that the balance can be set to read zero when the empty container is in place; thus, the measured mass with aspirin added is simply the mass of aspirin.

III. Determining Melting Points

One of the classic ways to identify and determine the purity of an organic chemical compound such as aspirin is to measure its melting point. Every pure chemical compound has a characteristic temperature at which it melts. The melting point also gives an indication of the purity because any impurities present will lower the melting point. Therefore, a melting point lower than the accepted value indicates either that the compound is impure or that it is not what you think it is. Pure compounds usually melt sharply within a range of 1 degree Celsius. Impure compounds melt over a much wider temperature range.

Your laboratory may have a special apparatus for determining the melting point of organic compounds, in which case your instructor will demonstrate its use. The instructions below are for a simple procedure using a large test tube filled with glycerol. *Figure 26.2* shows a similar apparatus but using a beaker rather than a test tube (and ethylene glycol instead of glycerol).

1. Obtain a thermometer, a notched rubber stopper, a large test tube half filled with glycerol, a small rubber band, and three melting-point capillary tubes that are closed at one end.

2. Your instructor will demonstrate the following procedure for adding a powdered material to a capillary tube. Tap the open end of a capillary tube into a small pile of purified aspirin, filling the tube with powder to a height of about 1 cm. Invert the tube so the closed end is down and gently tap the round sealed end on a bench top. This will cause the sample to settle and pack properly in the rounded end of the tube

3. Label the tube with a small piece of masking tape near the open end.

4. In the same fashion, fill another capillary with the salicylic acid that you started with and fill a third tube with the crude aspirin.

5. Place the tubes against the thermometer and hold them in place with a rubber band. The closed end of the tubes should be aligned with the middle of the thermometer bulb. The thermometer scale should be visible when looking at the tubes, and the rubber band should be high enough on the tubes so it will not be submerged in the glycerol (*Figure 26.2*).

6. Slip a split cork over the thermometer and assemble the apparatus as shown in *Figure 26.2*.

7. Use a microburner or a Bunsen burner with a very low flame to slowly heat the bottom of the test tube. The solids should melt between 80° and 160°C, so you can bring the temperature up fairly rapidly to about 80° but then let it increase slowly (only a few degrees a minute). Observe the contents of the melting-point tubes and record the temperature at which each of the samples just begins to melt and also the temperature at which they each completely melt. The key to obtaining accurate results is slow, careful heating. The thermometer heats up more slowly than the capillary tube, and slow heating ensures that it has time to "catch up" with the capillary tubes.

8. Record the melting temperatures in the data table.

Clean Up

Your instructor will tell you what to do with the samples of aspirin and will provide instructions for cleanup.

Figure 26.2 Diagram of melting-point setup

Questions To Be Answered After Completing This Experiment

Write out answers to the following questions on a separate sheet of paper and hand it in along with the data sheet.

Questions 1–6 can be answered while the purified aspirin is drying.

1. Draw the structures of salicylic acid and aspirin. Circle and identify the functional groups present in each compound.

2. The equation for the synthesis reaction is given in the Introduction. Copy this equation and indicate on the structures which bonds are broken in the reaction and how the resulting broken pieces pair up to give the products.

3. Sulfuric acid is a <u>catalyst</u> in this reaction. What does that mean?

4. What impurities do you think are probably in your crude aspirin product? Where do those impurities go in the recrystallization process?

5. What safety, environmental, and economic considerations would be important to a company manufacturing 1 million pounds of aspirin each year by this procedure? Which ones do you think are most important? Why?

6. The aspirin synthesized in this experiment should *never* be taken home for medicinal use. Why not?

Questions 7–9 should be answered after completion of the experiment.

7. Does your purified aspirin appear to be purer than crude aspirin? What is your evidence?

8. Impurities tend to lower the melting-point temperature of a solid and make it melt over a range of temperatures rather than at one sharp temperature. The published melting-point temperatures are 159°C for salicylic acid and 135°C for aspirin. Discuss your melting point data compared to these standard values.

9. You started the synthesis with 0.015 mole of salicylic acid.

   a. If the reaction worked perfectly, how many moles of aspirin would you make? (Look at the balanced equation for the synthesis reaction.)

   b. Convert this answer into grams of aspirin. To do this, multiply the number of moles of aspirin by the molar mass of aspirin (180 g/mole).

   c. Compare your actual mass of purified aspirin to the predicted mass from part 9.b. What is your percent yield?

   $$\text{percent yield} = \frac{\text{actual mass}}{\text{predicted mass}} \times 100\%$$

   d. Suggest two possible reasons why your percent yield is less than 100%.

Name _____   Date _____

Partner _____   Lab Section _____

Data Sheet—Experiment 26

Physical observations of salicylic acid:

Physical observations of the crude aspirin product:

Physical observations of the purified aspirin product:

Mass of salicylic acid used (grams):   _____

Mass of crude aspirin (grams):   _____

Mass of purified aspirin (grams):   _____

Melting point of salicylic acid: Melting starts:_____Melting complete:_____

Melting point of purified aspirin: Melting starts:_____Melting complete:_____

Melting point of crude aspirin: Melting starts:_____Melting complete:_____

How Much Fat Is in Potato Chips and Hot Dogs?

Introduction

Fat, sugar, and salt are three food components that worry nutritionists and provide copy for newspaper editors because Americans tend to consume too much of these food ingredients. In this and the next experiment, you will measure the amounts of these first two components in some foods that could be part of a typical fast-food meal.

Two approaches will be used to determine the fat content of foods. In fat-coated foods such as potato chips or french fries, the fat can simply be dissolved out of the food using a solvent. However, another approach must be used for meat products (such as hot dogs) because the fat is bound up in animal tissue (protein). Therefore, the meat must be chemically treated to break down the protein and other tissues in order to liberate the fat. Since the liberated fat can float on water, it can be easily removed and weighed.

I. Fat in Potato Chips

This study is conveniently done as a class project in which each student group analyzes a different brand of potato chips or other snack chips. Alternatively, each pair of students can do one measurement on each of two kinds of chips, then pool their results with those of other students. Your instructor will indicate how the class assignment will be structured. The essential procedure is very simple: Petroleum ether (a commercial mixture of hydrocarbons that is widely used as a solvent) is mixed with ground-up chips to extract the fat. After separating the solvent mixture from the chips, the petroleum ether is allowed to evaporate, leaving behind the fat that can then be weighed.

Overview of the Experiment, Part I

1. Obtain two weighed samples of potato chips or other snack chips.
2. Grind up the samples and mix them with petroleum ether to extract the fat.
3. Separate the petroleum ether from the potato chips by filtration into a weighed beaker.
4. Evaporate the petroleum ether on a steam bath or hot plate.
5. Weigh the beaker and calculate the mass of fat by difference.
6. Calculate the percent fat in the chips.

Materials Needed, Part I

Chemicals	Equipment
• 4–6 grams of chips • 50 mL of petroleum ether	• mortar and pestle • glass funnel and a funnel support • 2 sheets of filter paper to fit the funnel • 2 50-mL beakers • spatula • hot plate or steam bath

EXPERIMENTAL PROCEDURE – PART I

SAFETY NOTE:
Part I uses petroleum ether, a gasoline-like solvent that is *extremely* flammable. Therefore, it is absolutely essential that no open flames be present anywhere in the laboratory during this experiment. **Under no circumstances should the heating be done with a Bunsen burner.**

This procedure will be done twice, using two samples of chips. If there is sufficient equipment available, two students working as a team can do the two samples simultaneously.

1. Obtain two small beakers that are clean and dry. Label them with your initials and identifying numbers (1 and 2).

2. Take a bag of chips, the two beakers, and this lab manual or the data sheet to a laboratory balance. Accurately weigh out two samples of chips of about 2–3 grams each. This is done most conveniently by using small sheets of paper or small plastic weighing dishes. Record the mass of the first paper or dish to the nearest 0.01 gram, add 2–3 grams of chips, and again record the mass to the nearest 0.01 g. Repeat with the second sample. Keep track of which sample is which.

    Reminder about weighing: It is important to check the balance **each** time it is used to be sure it reads 0.00 g when there is nothing on the balance pan.

3. Also weigh the two beakers and record their masses to the nearest 0.01 gram in the appropriate places in the data table. (First check to be sure that the balance reads 0.00 g when empty.)

4. Put chip sample #1 in a clean porcelain mortar and use a pestle to crush the chips into very small pieces. (Because of the fat, the mixture may be rather sticky.)

5. Add 15 mL of petroleum ether to the mortar and grind the mixture thoroughly.

6. Prepare a glass funnel with folded filter paper (as shown by the instructor). Mount it over weighed beaker #1. Using a spatula, carefully transfer the potato-chip mixture into the filter. Try to get most of the mixture into the filter.

7. In order to rinse out any remaining fat in the mortar and on the chip mixture, add 5 mL more petroleum ether to the mortar, stir it around with the pestle, then pour it onto the chip mixture in the filter.

8. Repeat with another 5-mL rinse of the mortar and the chip mixture in the filter.

9. Repeat steps 4–8 with the second sample of chips using the same mortar and pestle but a fresh piece of filter paper and weighed beaker #2.

10. When both mixtures have finished filtering, the next step is to remove all of the petroleum ether by evaporation. There are two ways this can be done.

    a. You can leave the beakers in a fume hood until the next day or the next lab period, by which time the solvent will have evaporated.

    b. A much faster way is to place the beakers on a steam bath or hot plate <u>in a fume hood</u>. Leave them for about 15 minutes or until all of the petroleum ether has evaporated. Then remove the beakers from the steam bath or hot plate, carefully wipe the outsides to remove all water, and let them cool for a few minutes.

11. Reweigh the beakers and record the masses to the nearest 0.01 gram. (First check to be sure the balance reads zero when empty.)

Clean Up

Discard the filter paper and chip mixtures in a designated container. The instructor will specify what to do with the mortar and pestle (which should be nearly clean) and the beakers (which are quite greasy).

Calculations for Part I

For each sample, calculate by subtraction the mass of chips and the mass of fat. Then calculate the percent fat in the chips:

$$\% \text{ fat } = \frac{\text{mass of fat}}{\text{mass of chips}} \text{ x } 100 \tag{1}$$

Sharing Your Results

Finally, share your results with others in your class or lab section. It will probably be advantageous to report your two results separately rather than as an average. If others analyzed the same brand, you can see how closely you agree. Examining the class data as a whole, you can look for differences and possible generalizations about different types of snack chips.

II. Fat in a Hot Dog (or other meat sample)

This procedure is written for hot dogs, but it can be used for any kind of meat sample. Your instructor may divide up your class and have different student groups analyze various hot dog samples. For example, your class may wish to investigate whether different brands of hot dogs differ significantly in fat content, whether chicken or turkey hot dogs differ from beef and pork hot dogs, whether hot dogs and hamburgers differ significantly in fat, whether "low fat" products are really as claimed, and how vegetarian hot dogs compare to meat hot dogs in their fat content.

The experimental procedure utilizes a special solution identified simply as "protein liquefying reagent." This is a highly alkaline mixture containing sodium hydroxide, sodium salicylate, potassium sulfite, isopropyl alcohol, and water.

Overview of the Experiment, Part II

1. Obtain a sample of ground hot dog or other meat.
2. Determine the masses of two separate hot-dog samples in test tubes.
3. React the hot-dog meat with protein liquefying reagent.
4. Cool and centrifuge the samples.
5. Remove the fat layer from each hot-dog sample and weigh it.
6. Calculate the percent fat in a hot dog.

Materials Needed, Part II

Chemicals	Equipment
• 4 grams of ground hot-dog meat or other kind of meat (several brands may be available) • 10 mL of protein liquefying reagent	• hot plate • 2 15-mL centrifuge tubes • 250-mL beaker • 2 small vials or other containers • 2 long-stem glass Pasteur pipets • pipet bulb • thermometer • centrifuge

EXPERIMENTAL PROCEDURE – PART II

SAFETY NOTES
- Do not allow the protein liquefying reagent (step 4) to contact your skin. It is highly caustic and can cause serious skin damage.
- Check that there are no open flames near you as you work since the protein liquefying reagent is flammable. Heat the samples ONLY on a hot plate. Do NOT use an open flame. The vapor from the protein liquefying reagent is extremely flammable.

---

1. Put approximately 75 mL of water in a 250-mL beaker. Place the beaker of water on a hot plate and heat the water to between 80° and 90°C.

2. While the water is heating, obtain two centrifuge tubes and label them #1 and #2. Put the labels high on the tubes near the top. Take the tubes along with the lab book or the data sheet to a laboratory balance. Check to be sure that the empty balance reads exactly 0.00 g, then weigh the test tubes and record their masses to the nearest 0.01 gram.

3. Put between 2 and 2.5 grams of ground hot-dog meat in each test tube (your instructor will show you about how much this is) and reweigh the tubes. Record the mass of each tube + hot-dog sample to the nearest 0.01 gram.

4. Add about 5 mL of the protein liquefying reagent to each hot-dog meat sample.

5. Put the test tubes in the beaker of water and heat the beaker **on a hot plate** until the reagent in the tube starts to boil (about 80°C). Maintain the temperature of the water so that the contents of the test tubes boil for 10 minutes. Do not leave the beaker and test tubes unattended. The liquid could boil out of the test tube.

6. After the mixture has boiled for 10 minutes, it should be dark brown with some yellow fat floating at the top. Remove the test tubes from the hot water, stand them in a 250-mL beaker, and let them stand until they are cool enough to handle. Do not let them cool completely because the fat may turn from liquid to solid.

7. Label two small containers (vials, small beakers, test tubes, or watch glasses) as "sample 1" and "sample 2" and then weigh them to the nearest 0.01 gram. (Remember to check first to be sure the balance reads 0.00 g when empty.) Record the masses on the data sheet.

8. Put the two test tubes in a centrifuge, placing them directly opposite each other. Other students may add pairs of test tubes at the same time. Centrifuge the mixtures to completely separate the fat from the rest of the solution. The fat should be floating on top.

9. Use a Pasteur pipet (glass tube with a long, thin stem and a small rubber bulb attached to the top) to slowly transfer the top fat layer from test tube #1 to the correct preweighed container. Be very careful to remove all of the fat but none of the brown liquid. Work slowly; it takes patience to do this correctly so that only the fat is removed. If you accidentally suck up some of the brown liquid, it can usually be removed from the bottom of the container of fat by using the Pasteur pipet.

10. Repeat the procedure for the other sample.

11. Weigh the containers of fat to the nearest 0.01 gram and record their masses in the data table.

Clean Up

The "brown protein liquid" remaining in the test tubes can be washed down the drain with plenty of water. Your instructor will tell you what to do with the fat in the small vials.

Calculations for Part II

By subtraction of masses in the data table, calculate the mass of hot dog and the mass of fat for each sample. The percent fat is then calculated from the relationship:

$$\% \text{ fat} = \frac{\text{mass of fat}}{\text{mass of hot dog}} \times 100$$

Sharing Your Results

This is a fairly time-consuming analysis, so you were probably limited to doing two replicate samples of one meat product.   Other members of your class probably have results for other meat products. When you are finished with the calculations, share your results with the rest of the class or lab section. If different brands of hot dogs were analyzed, how do the fat contents compare? Can you draw any conclusions about relative fat content of different kinds of foods?

Questions To Be Answered After Completing This Experiment

Write out answers to the following questions on a separate sheet of paper and hand it in along with the data sheet.

Part I: Fat in Potato Chips

1. Looking at the class results, can you make any generalizations? For example, do potato chips have more or less fat than tortilla chips?

2. If your class analyzed any "low-fat" or "no-fat" products, such as WOW™ potato chips, do the results support this claim? If not, suggest a reasonable explanation.

3. Potatoes are naturally low in fat, so where does the fat in potato chips come from? How are low-fat chips different?

4. A snack pack of potato chips or other kinds of chips holds 1 ounce (28 grams) of chips. Based on your data, how much fat is present in a snack pack? In such a bag of chips, there are about 15 grams of carbohydrate and about 1 gram of protein. Given that carbohydrates and protein provide about 4 Calories of energy per gram and fats provide about 9 Calories per gram, what is the total energy equivalent (in Calories) of a 1-ounce bag of chips? What percent of the Calories are from the fat? *Show your calculations.*

Part II: Fat in Hot Dogs

1. If your class analyzed different kinds of hot dogs or other meats, what can you conclude about the fat content in these products? Are some products significantly lower in fat than others?

2. Why did hot dogs require a different fat extraction procedure than the chips did? Which of the two procedures do you think would work best when analyzing the following foods: roasted peanuts, corn, pepperoni pizza? Explain your reasoning for each food.

3. A typical hot dog weighs about 45 grams. Using your data, calculate the number of grams of fat in the hot dog you analyzed? There are about 7 grams of protein in a hot dog. Almost all of the remaining mass is water. Calculate how many Calories are provided by the fat, by the protein, and by the water. What percentage of the Calories in the hot dog can be attributed to the fat?

Name _____   Date _____

Partner _____   Lab Section _____

Data Tables–Experiment 27

Part I: Percent Fat in Potato Chips or Tortilla Chips

Brand of chips _____

	Sample 1	Sample 2
Mass of dish + chips		
Mass of empty dish		
Mass of chips		
Mass of beaker + fat		
Mass of empty beaker		
Mass of fat		
Percent fat in chips		

Part II: Percent Fat in a Hot Dog (or other meat product)

Description of the sample: _____

	Sample 1	Sample 2
Mass of test tube + hot dog		
Mass of empty test tube		
Mass of hot dog		
Mass of container + fat		
Mass of empty container		
Mass of fat		
Percent fat in hot dog		

How Much Sugar Is in Soft Drinks and Fruit Juices?

Introduction

Fat, sugar, and salt are three food components that most Americans consume in excess. For example, they are present in large amounts in typical fast-food meals. Experiment 27 focused on fat in potato chips and hot dogs. This experiment focuses on sugar in beverages. Almost all soft drinks that are not milk-based (Coke, Pepsi, Sprite, Kool-aid, Gatorade, etc.) are essentially sugar solutions with small amounts of additives for flavoring and color. Fruit juices are also mostly sugar solutions with small amounts of other materials. As you will see, most of these beverages contain a surprisingly large amount of sugar.

Background Information

In this experiment, you will determine the sugar content of various beverages by measuring the density of each beverage, using a simple hydrometer. When sugar is dissolved in water, the density of the resulting solution increases. By measuring this density increase for various beverages and comparing it with sugar solutions of known concentrations, it is possible to determine the approximate concentration of sugar in the beverages. **Density** is defined as the mass (in grams) of 1 cubic centimeter, or 1 milliliter, of a substance (see Section 5.8 in *Chemistry in Context*, fourth edition).

A hydrometer is a device designed to quickly measure the density of a liquid. It is based on the fact that floating objects displace their weight in liquid. A less-dense floating object displaces less liquid when it floats than a more-dense floating object. Therefore, the less-dense object floats higher in the liquid. When the density of the liquid increases, objects float higher because a smaller volume of liquid needs to be displaced to achieve the same weight of liquid. (One interesting example of this is that you float higher in salt water than you do in fresh water.) A hydrometer is simply a floating object with a scale to measure how high it floats in various liquids.

Overview of the Experiment

1. Prepare a hydrometer using a thin-stem plastic pipet.
2. Calibrate your hydrometer using four sugar solutions of known concentration.
3. Use your data to prepare a graph of hydrometer depth versus sugar concentration.
4. Use your hydrometer to investigate several soft-drink samples.
5. Use your graph to determine the sugar concentration in each soft-drink sample.

Materials Needed

Chemicals	Equipment
• 50 mL each of four sugar solutions: 4% sugar, 8% sugar, 12% sugar, and 16% sugar (wt/volume %) • 50-mL samples of different soft drinks, previously decarbonated	• 50-mL graduated cylinder • thin-stem plastic pipet • 1 1/2 inch nail • metal nut, 3/16 inch or 8-32 size • ruler graduated in millimeters

EXPERIMENTAL PROCEDURE

I. Constructing a Hydrometer

You will construct a hydrometer of the type shown in *Figure 28.1*.

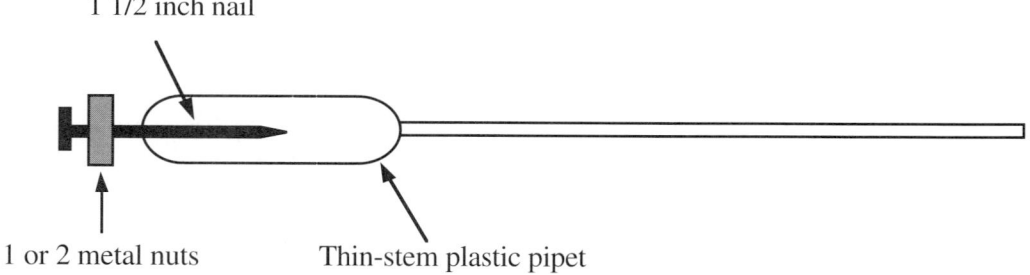

1 1/2 inch nail

1 or 2 metal nuts   Thin-stem plastic pipet

Figure 28.1 A simple homemade hydrometer

1. You will need a thin-stem plastic pipet, a new 1 inch or 1 1/2 inch nail with a large head, and one or two metal nuts of the type to fit on machine screws.

2. Put the nut(s) on the nail and then *gently* push the nail through the center of the bottom of the bulb of the pipet as shown in *Figure 28.1*. After the nail is in place, do not move it. If you do, the pipet may leak. If the nail did not go in straight, start again with a new pipet.

3. Fill the pipet approximately half-full of water.

4. Place the pipet, bulb end down, into a 50-mL graduated cylinder filled nearly to the top with water at room temperature. You need to adjust the amount of water in the pipet so that it floats with the bulb *near* the bottom (but not touching the bottom) and with only a short length of the stem sticking out of the water (as shown in the *Figure 28.2*). If it floats too high or too low, adjust the height by cautiously adding or removing some water from the bulb of the pipet. (Your instructor may give a demonstration of how to do this.) When you have the pipet floating correctly, it is ready to use as a hydrometer.

If the hydrometer is placed in a liquid that is denser than water, it will float with more of its stem out of the water. To determine the relative density, and hence the amount of dissolved solids (mostly sugar) in soft drinks or fruit juices, all you need to do is float the hydrometer in the beverage and measure the length of the stem that sticks out of the liquid. The longer the protruding stem, the denser the liquid, and the greater the sugar content. The reference point for the protruding height will always be the surface of the liquid.

Figure 28.2 The hydrometer assembly

II. Calibrating the Hydrometer

In order to estimate the actual amount of sugar in a soft drink, the hydrometer must be calibrated.

1. Start with the graduated cylinder containing pure water. This is the 0% sugar reference solution.

2. Place the hydrometer gently in the cylinder as shown in *Figure 28.2*.

3. Make sure the hydrometer is free floating. To do this, tap it gently once or twice to be sure that it bobs freely and that it does not stick to the sides of the container.

4. Using a ruler, carefully measure the height of the hydrometer stem that sticks out of the fluid, recorded to the nearest millimeter (mm). Tap the hydrometer gently and repeat the measurement. This measurement is a little tricky since it must be from the liquid surface to the exact end of the hydrometer stem. The height variations between liquids will be small, so it is important to measure the heights carefully. (If you are working with a partner, both of you should check the height.) Record the height in the data table.

5. A set of graduated cylinders filled with reference sugar solutions (4%, 8%, 12%, and 16%) will be available at labeled stations in the laboratory. You will be making measurements with your hydrometer in each solution, but the *order* in which you do them does not matter.

6. Take your hydrometer to one of the sugar solutions. Gently blot water off of the hydrometer and place it in the sugar solution. Using the same procedure as described previously, measure the height of the protruding hydrometer stem for the sugar solution. Record this number in the data table.

7. Proceed to make similar measurements for the other three sugar solutions. *To avoid contaminating the solutions, you should always rinse your hydrometer with pure water and blot it gently before inserting in a new solution.* Record the heights in the data table.

III. Testing Soft Drinks and Other Beverages

Important note: Carbonated drinks must first be "decarbonated" because if the hydrometer is placed in a carbonated drink, bubbles sticking to the hydrometer will add buoyancy and the reading will be incorrect. The carbonated beverage samples that you will be testing have been decarbonated in advance by boiling for a few minutes and then cooling to room temperature.

Your instructor will specify what samples to test, or you may be offered a variety of samples from which to choose. Members of your class or lab section may be asked to test different samples so that class data can be assembled and compared. The beverage samples will be in graduated cylinders at labeled stations in the lab.

> **Safety Note**: Do not drink any beverages that have been opened or used in the laboratory.

1. Using the procedure described in Part II, place your hydrometer in each of the beverage samples and carefully measure the exact height of the protruding stem. *To avoid contaminating the samples, always rinse the hydrometer with pure water and blot it gently before placing it in a new solution.*

2. Record your data in the data table.

Optional: If your instructor permits this, you may take your hydrometer with you (being careful not to squeeze out any of the water) and use it to test other drinks you have in your house or residence hall.

> **Caution**: Do these tests only on small portions of beverages which should then be discarded. DO NOT DRINK ANY SAMPLES IN WHICH THE HYDROMETER HAS BEEN IMMERSED.

IV. Calculations

The way to make the calibration data more useful is to plot the data as a graph, plotting the hydrometer height (on the vertical axis) against % sugar (on the horizontal axis). You can use a sheet or graph paper or the graph that is provided on the data sheet, which has the horizontal axis already labeled. The vertical axis should be labeled as "height in millimeters." You will need to think carefully about the scales for the vertical axis so that the graph extends over a large part of the paper. *Note that it is NOT necessary for the vertical scale to start at zero.* Enter the data points for the calibration solutions from Part II on the graph (each point has a % sugar value and a height value) by placing a dot in the correct location and drawing a small circle around it so that it is clearly visible. You should have five points on the graph. Connect the points with a smooth line. The line is not necessarily straight; it may be curved, but it should not be drawn as a zig-zag. Therefore you should examine the points carefully before drawing a line.

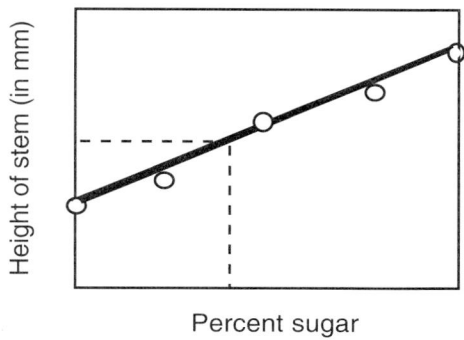

Figure 28.3 Sugar calibration graph

To determine the percent sugar in a soft drink, locate the hydrometer height for that sample on the vertical scale and draw a horizontal line until it intersects the calibration line, as shown above (*Figure 28.3*). From that point, draw a vertical line down the horizontal axis and read the percent sugar from the horizontal scale.

Optional: Use of a computer for graphing. If a computer with a graphing program is available, the data can be plotted, and then the equation for the line can be obtained. With this equation, it is a simple matter to calculate the % sugar from the hydrometer height for each of the beverages.

Sharing Your Results

When the calculations are completed, share your results with other members of the class or lab section. Try to assess the degree of agreement between different student results for the same samples. See what conclusions you can draw about different kinds of beverages.

Questions To Be Answered After Completing This Experiment

Write out answers to the following questions on a separate sheet of paper and hand it in along with the data sheet.

---

1. Which beverages tested by you or your classmates had the highest sugar content? Were there any surprises in the results for sugar content in beverages? If so, what were they?

2. Do the experimental results for "sugar-free" beverages support the manufacturer's claim of no sugar?

3. If fruit juices were tested, how does the sugar content compare to typical colas and carbonated beverages?

4. How could you modify your hydrometer to increase its accuracy?

5. A bottle of regular Pepsi states the following on its label:
   Serving size: 8 fluid ounces (about 240 mL)
   Total carbohydrate (all as sugar): 27 g

   a. Calculate to the nearest gram the number of grams of sugar you will consume if you drink a 20-ounce bottle of Pepsi.

   b. If a teaspoon of sugar is approximately 4 grams, how many teaspoons of sugar would you have consumed from this bottle of Pepsi?

6. Suppose you wished to find the percent of alcohol in an alcoholic beverage. Alcohol has a density of 0.79 gram/mL and water has a density of 1.00 gram/mL. Describe how you would use a hydrometer to accomplish the task. What modification might you have to make in your hydrometer?

7. When cans of regular and diet soda are added to a large container of ice water, some of the cans float while others sink. Which float? Which sink? Explain why.

Name _____   Date _____

Partner _____   Lab Section _____

Data Sheet–Experiment 28

Part I: Hydrometer Calibration

Percent sugar	0%	4%	8%	12%	16%
Height of hydrometer stem (in mm)					

Part II: Tests of Beverages

Beverage tested				
Height of hydrometer stem (in mm)				
Percent sugar				

(If more beverages are tested, give the results on the back of this sheet.)

Percent sugar

Vitamin C in Fruit Juice and in a Vitamin Tablet

Introduction

Vitamin C is one of the essential vitamins required for good health. Most people obtain the necessary amount of vitamin C in their diet (primarily in fresh fruits, fruit juices, and vegetables), but vitamin C tablets are also widely used as nutritional supplements. In this experiment, you will analyze a sample of fruit juice to find out how much vitamin C it contains and see whether it could provide your daily requirement for vitamin C. You will also analyze a vitamin C tablet.

Background Information

Vitamin C, or ascorbic acid ($C_6H_8O_6$), is chemically similar to the simple sugar, glucose ($C_6H_{12}O_6$), that is plentiful in our bodies. Most animals possess an enzyme needed for making ascorbic acid from glucose, but humans and a few other species lack that enzyme and therefore must secure ascorbic acid directly from foods they consume. The best-known function of ascorbic acid is the prevention of scurvy; a minute amount (10 mg per day) is adequate for this purpose. However, it also plays several other important roles in human health. Most of its functions are related to the fact that it has a strong tendency to transfer electrons to other chemical substances (chemists say that it is a "reducing agent" or "antioxidant"). It acts to prevent other, potentially harmful, reactions that involve the transfer of electrons.

To maintain good health, the current recommended daily allowance (U.S. RDA) for vitamin C is 60 mg. Foods like fruits and vegetables with a high water content often contain large amounts of vitamin C. Vitamin C is very soluble in water, which means that, if high doses of vitamin C are ingested, much of it is rapidly excreted in urine.

The **titration method** of analysis has been explained and utilized in several previous experiments and will not be described in detail here. You may wish to refer back to your previous experience with titrations. The titration analysis for ascorbic acid (vitamin C) is based on its tendency to lose electrons. A suitable electron acceptor substance, in this case iodine (I_2), is used as the reagent to "titrate" the ascorbic acid. Although it is not necessary to understand the reaction in order to do the analysis, the chemical equation for the reaction is the following:

$$C_6H_8O_6 + I_2 \rightarrow C_6H_6O_6 + 2\ I^- + 2\ H^+$$

Since it is difficult to prepare iodine solutions with an accurately known concentration, the iodine concentration is determined first by titrating a reference sample of ascorbic acid. The "end point" of the titration is signaled by the appearance of the color of unreacted iodine. The color is made more intense by adding a drop of starch solution, which forms a deep blue color with iodine.

Overview of the Experiment

1. Assemble all of the necessary materials.
2. Titrate the ascorbic acid reference solution.
3. Prepare a solution of a vitamin C tablet in a volumetric flask.
4. Titrate portions of the solution from the vitamin C tablet.
5. Calculate the amount of ascorbic acid in the vitamin tablet.
6. Titrate orange juice or other fruit juice.
7. Calculate the concentration of ascorbic acid in the juice.

Materials Needed

Chemicals	Equipment
• vitamin C tablet, 100 mg • ascorbic acid reference solution • iodine solution • starch solution • distilled water • orange juice or other fruit juice	• 24-well wellplate • small stirring rod • 2 graduated-stem plastic pipets • 100-ml volumetric flask with cap • 2 50-ml beakers (or other small containers)

Place a piece of white paper under the wellplate so that it will be easier to see the color changes as you do the experiment.

EXPERIMENTAL PROCEDURE

I. Preparing the Vitamin C Tablet Sample

1. Weigh a vitamin C tablet and record its mass to the nearest milligram (0.001 gram).

2. Place the tablet in a clean 100-ml volumetric flask. Fill the flask about half-full with water. Shake and swirl the flask until the tablet is broken down. Label the flask and set it aside to continue dissolving while you go on to the next procedure.

II. Titration of the Ascorbic Acid Reference Solution

1. Using a clean, dry beaker, obtain about 10 ml of the ascorbic acid reference solution. Record its exact concentration (given in mg/ml).

2. Fill a graduated-stem plastic pipet with exactly 1 ml of the reference ascorbic acid solution. (It may take a little practice to fill the pipet exactly to the 1-ml mark.) Then slowly dispense the 1-ml sample into a clean, dry well in the wellplate (e.g., well A1).

3. In the same fashion, add exactly 1 mL of ascorbic acid reference solution to each of 3 or more wells.

4. Add 1 drop of starch solution to each of the four wells.

5. Obtain a small supply of iodine solution in a clean, dry beaker or test tube.

6. Fill the other pipet with iodine solution.

7. Proceed to titrate the first ascorbic acid reference sample with the iodine solution, counting drops and stirring as you go. You will observe a momentary blue color that disappears upon stirring but lingers longer as more iodine is added. The end point occurs when one drop gives a pale blue color that does not disappear upon stirring. Record your data on the data sheet.

   - You may wish to designate one well in the wellplate for waste. Partial drops can be dispensed into it.

   - Remember that the most consistent drop sizes are obtained by holding the pipet vertically.

   - If you miss the endpoint or lose count of drops, simply go on to another sample. It is easy and fast to titrate additional portions of the solution as needed.

8. In the same fashion, titrate at least 3 more ascorbic acid reference samples. Record the results on the data sheet.

III. Titration of the Vitamin Tablet Solution

1. By now, the vitamin C tablet in the volumetric flask should be mostly dissolved (except for insoluble "filler" material that is typically starch).

2. Add deionized water to fill the flask up to the bottom of the neck, then *slowly* add more water to bring the water level up the line etched on the neck of the flask. The last milliliter or so should be added one drop at a time. The bottom of the curved meniscus should just touch the line as shown in the following figure.

Water surface ("meniscus")

Etched line

Neck of volumetric flask

3. Put the cap on the flask securely and mix the contents of the flask thoroughly by repeatedly turning the flask upside down and swirling. (If necessary, hold the cap with a finger.)

4. Take the graduated-stem pipet that was used for the reference solution. Rinse it thoroughly with deionized water (filling and emptying several times) and then rinse it with the solution from the volumetric flask.

5. Use this pipet to add exactly 1 ml of the vitamin tablet solution to each of four clean wells in the well plate (e.g., row B).

6. Add a drop of starch solution to each of these samples.

7. Using the same iodine solution and plastic pipet that you used in Part II, titrate the first sample to the pale blue endpoint. Record the result.

8. Repeat the titration of the vitamin C tablet solution at least 3 more times and record the results.

IV. Titration of a Juice Sample

> **SAFETY NOTE:** Do not consume any beverage from a container that has been open in the laboratory.

1. Rinse thoroughly with deionized water the beaker used in Part II. Also rinse the plastic pipet thoroughly or obtain a new pipet for the juice sample.

2. Wipe the beaker dry; then obtain a sample of orange juice or other juice. If several juice samples are being tested, record which one you used.

3. Rinse the pipet with the juice sample.

4. Use this pipet to add exactly 1 ml of juice to a fresh well in the wellplate.

5. Add one drop of starch solution.

6. Titrate this juice sample in the same way you titrated the vitamin C tablet solution. The color of the juice affects the appearance of the end point; therefore, it is good idea to do one titration for practice to see what the end point looks like. Depending on the number of drops used, you might choose to use a larger or smaller volume of juice for the subsequent titrations. If so, be sure to record the volumes actually used.

7. Then do at least 3 more titrations of the juice. Remember to add a drop of starch solution to each. Record your data.

V. Optional Extensions or Alternate Assignments

A variety of juices and foods can be analyzed for vitamin C by this basic method. Once you know how to do this analysis and you have done the calibration with a known reference solution, it is easy to analyze additional samples. Your instructor may divide up the class so that different samples will be analyzed by various members of the class. In addition to fruit juices and other beverages, it is possible to extract the liquid from vegetables such as broccoli or peppers (both of which are high in vitamin C) and titrate the extracted liquid.

If you analyze additional samples, record the data on a separate sheet of paper, using the same format as shown on the data sheet.

VI. Calculations

1. Look carefully at the data from the three sets of titrations to see whether the results show consistency or whether any individual result should be eliminated because it appears to be very different from the other data in the set. If so, make a note beside that value and omit it from your calculations.

2   Calculate the average number of drops of iodine used for each set of titrations.

3. For the first set, divide the milligrams of ascorbic acid per milliliter by the average number of drops of iodine used (equation 1). This gives the mg of ascorbic acid corresponding to 1 drop

of iodine solution. This is the **calibration factor** for this analysis. It is unique for the particular iodine solution and your equipment.

$$\frac{\text{mg ascorbic acid}}{1 \text{ mL of ref. sol'n.}} \times \frac{1 \text{ mL of ref. sol'n.}}{\text{drops of iodine}} = \frac{\text{mg ascorbic acid}}{1 \text{ drop of iodine}} \qquad (1)$$

4. For the analysis of the vitamin C tablet, multiply the average number of drops of iodine used by the calibration factor to obtain the mg of ascorbic acid per mL (equation 2).

$$\frac{\text{drops of iodine}}{1 \text{ mL sample}} \times \frac{\text{mg ascorbic acid}}{1 \text{ drop of iodine}} = \frac{\text{mg ascorbic acid}}{1 \text{ mL sample}} \qquad (2)$$

Multiply the answer from equation (2) by 100 to find the total milligrams of ascorbic acid in the 100-mL volumetric flask. This is the same as the milligrams of ascorbic acid in the original tablet.

5. For the analysis of the juice sample, use equation (2) to calculate the milligrams of ascorbic acid per milliliter of juice.

6. If a standard serving of juice is 8 ounces and 1 fluid ounce is 30 mL, calculate how much vitamin C would be obtained from an 8-ounce serving of this juice.

Questions To Be Answered After Completing This Experiment

Write out answers to the following questions on a separate sheet of paper and hand it in along with the data sheet.

---

1. Is vitamin C a water-soluble or fat-soluble vitamin? How does this relate to the procedure we used for analysis?

2. What fraction of the mass of the vitamin tablet was ascorbic acid?

3. How closely did your analysis of the vitamin tablet match the label on the bottle? Calculate the percentage difference between them.

4. The current (2002) U.S. RDA (recommended dietary allowance) for vitamin C is 75 mg/day for young adult women and 90 mg/day for young adult men. How much of the juice that you analyzed would you have to drink to meet the U.S. RDA for vitamin C? (**Note:** some nutritionists think the RDAs for vitamin C are too low and that 200 mg/day is more nearly optimal.)

5. If you analyzed other foods, how much of each would you have to eat to meet the U.S. RDA for vitamin C?

Notes and Calculations

Name _____ Date _____

Partner _____ Lab Section _____

Data Sheet—Experiment 29

Part II: Titrations of the Reference Solution

Concentration of reference ascorbic acid _____ mg/mL

Trial	#1	#2	#3	#4	#5
mL of reference sol'n	1.00	1.00	1.00	1.00	
Drops of iodine					

Note: *Extra columns are for optional additional titrations.*

Average drops of iodine solution = _____ per 1 mL of reference solution

Calibration factor = _____ mg of ascorbic acid per drop of iodine

Part III: Titration of Vitamin Tablet

Mass of tablet: _____

Titration	#1	#2	#3	#4	#5
mL of sample	1.00				
Drops of iodine					

Average drops of iodine solution per 1 mL of sample: _____

mg of ascorbic acid per 1 mL:_____ mg of ascorbic acid per tablet: _____

Part IV: Titration of a Juice Sample

Description of the sample: _____

Titration	#1	#2	#3	#4	#5
mL of sample	1.00				
Drops of iodine					

Average drops of iodine solution per 1 mL of juice: _____

mg of ascorbic acid per 1 mL: _____ mg of ascorbic acid per 8 ounces:_____

Isolation of DNA (Deoxyribonucleic Acid)

Introduction

Everyone has heard of DNA. It has been in the news frequently in recent years in conjunction with topics as diverse as criminal investigations, genetic engineering, and genetically induced diseases. Chapter 12 in *Chemistry in Context,* fourth edition, discusses DNA in considerable detail. In this experiment, you will be able to isolate and actually see some DNA. It will not be human DNA; instead it will be DNA from onions. Onion DNA may sound less exciting, but the same basic techniques are involved as those required for studies with human DNA.

Background

Deoxyribonucleic acid (DNA) molecules contain the blueprint for life and are found in the nucleus of any living cell. The makeup of DNA molecules is both simple and complex. It is simple because the molecule that contains the blueprint for a living organism is a polymer consisting of only six chemical fragments (phosphate, deoxyribose, guanine, cytosine, adenine, thymine). It is complex because the phosphate and deoxyribose bond together to form a framework to which the four other fragments are attached. The sequence of these four fragments on the framework is the basis for an organism's genetic code. A single strand of DNA can have thousands of fragments hooked together in a very specific order. Hydrogen bonding causes the long molecules to assume a helical shape. The helix consists of two strands of DNA wrapped around each other. (See Figure 12.7 in *Chemisty in Context*, fourth edition for a diagram of the double helix.) Many methods have been developed for extracting DNA from cells. The procedures are designed to extract only the DNA and leave other cell components behind. If you consider the number and types of proteins and lipids found in a cell, it is surprising that DNA is relatively easy to isolate, at least from some kinds of tissue. For example, the DNA in onion cells can be easily isolated by first disrupting the cell walls with a homogenizing solution consisting of detergent, salt, and a reagent to control the pH. The DNA that "leaks" out of the broken cells can be "precipitated" (made insoluble) by addition of ethanol. Because DNA molecules are long polymer chains, they tend to bind together when they are precipitated to form long strands that will mat together and look something like coagulated egg white. The strands can be wrapped around a thin glass or plastic rod and removed.

Overview of the Experiment

1. Weigh a sample of chopped onion.
2. Add homogenizing solution and incubate at 60°C.
3. Mash the onion preparation to break up cells.
4. Chill the mixture, then filter it to remove the cell walls.
5. Precipitate the DNA with ethanol.
6. Spool the DNA around a glass or plastic rod.

Materials Needed

Chemicals	Equipment
• chopped onion • 95% ethanol • 20 mL homogenizing solution • crushed ice	• 50-mL beaker • 2 large test tubes 150 x 30 mm • 50-mL graduated cylinder • thermometer • 55-mm funnel • coarse filter paper • mortar and pestle • small petri dish size or other flat dish • Pasteur pipet Shared equipment • laboratory balance • 60°C water bath

EXPERIMENTAL PROCEDURE

1. Weigh 10 grams of chopped onion into a large test tube.

2. Add 20 mL of "homogenizing solution" (see Background section) and place the test tube in a 60°C water bath for 10 minutes.

> **Note:** If a large temperature-controlled water bath is not available, half fill a 500-mL beaker with water and heat it to 60°C on a hot plate. Monitor the temperature with a thermometer. Place the test tube with the onion mixture in the beaker and maintain the temperature of the water bath at 60°C by adjusting the hot plate.

3. Keep the onion mixture in the beaker of 60°C water for 10 minutes. It is important to monitor the temperature and be sure that it stays close to 60°C. If the temperature of the onion mixture reaches 70°C, the DNA structure will be destroyed, and you will not be able to isolate and spool it. (**Note:** Incubating with the homogenizing solution serves to soften the cell walls, break down the protein in the cells, and may also help to dissolve cell membranes.)

4. After the onion mixture has heated for 10 minutes, put the test tube in an ice-water bath and chill until the mixture reaches 15°C.

5. Pour the cooled solution into a mortar and use the pestle to mash the onion to a smooth paste.

6. Transfer the paste and all the liquid to a 50-mL beaker and chill in an ice bath for 15 minutes.

7. Using a funnel, filter the mixture through coarse filter paper into a 50-mL graduated cylinder and return the liquid to the ice bath for 5 minutes.

8. Slowly pour 20 mL of 95% ethanol down the side of the graduated cylinder to form a layer on top of the DNA-containing liquid. A white stringy precipitate should form at the liquid/liquid interface.

9. Slowly pour the liquid into a small petri dish. Then put the end of a thin glass rod or a glass Pasteur pipet into the liquid and slowly twirl it around. White strands of DNA should wrap themselves around the end of the pipet.

Questions To Be Answered After Completing This Experiment

Write out answers to the following questions on a separate sheet of paper and hand it in along with the data sheet.

---

1. In this experiment, you were cautioned not to allow the temperature to go above 60°C. Why do you think DNA decomposes when it is heated to too high a temperature? (**Hint:** Think about the double-helix structure.)

2. DNA was precipitated out of a water solution by adding ethanol (CH_3-CH_2-OH). Why do you think DNA is less soluble in ethanol than in water? Remember that ethanol is partly a hydrocarbon and therefore is a less polar molecule than water.

3. Below are the structural formulas for the four nucleotide bases in DNA: adenine, thymine, cytosine, and guanine. What features do they have in common?

Adenine   Thymine   Cytosine   Guanine

4. Which hydrogen atoms in the 4 nucleotides are responsible for forming the hydrogen bonds that hold DNA strands together in the double helix? How <u>many</u> hydrogen bonds are possible for each? Suggest an explanation for why adeinine always pairs with thymine and cytosine always pairs with guanine.

5. In light of your answer to Question 4, suggest why DNA molecules that are rich in cytosine and guanine will "melt" or decompose at a higher temperature than DNA that is rich in adenine and thymine.

6. The diversity of life depends on the fact that a great many combinations of the 4 nucleotide bases (shown above) can be assembled into long chains. Illustrate this by writing out 20 of the 64 possible combinations of just 3 bases (instead of the thousands of bases in real DNA molecules), using any combinations of the 4 building blocks (designated as A, T, C, and G).

Name _____   Date _____

Partner _____   Lab Section _____

Data Sheet—Experiment 30

Describe the appearance of the DNA you isolated.

Performance-Based Assessment Activities

General Instructions: Your instructor may assign one or more of these activities as a test of whether you have learned how to carry out scientific investigations. First, write up a brief description of your experimental plan and check it with your instructor for feasibility and safety; then perform the experiment. You will need to keep a careful written record of your method and your observations. Upon completion of the experiment, you will be asked to describe your procedure and explain how your conclusions are based on your recorded observations.

1. A spectrophotometric study to follow Experiment 6

    Design and test a method for determining the relationship between the concentration of a colored solution and its absorption of light. Include in your procedure a way to determine the concentration of an unknown solution.

2. A quantitative chemical mole study to follow Experiment 9

    When calcium carbonate reacts with hydrochloric acid, there are three possible products: $CaCl_2$, $CaCl_2 \cdot 2H_2O$, and $CaCl_2 \cdot 6H_2O$. Design and carry out an experiment to find out which product is formed.

3. A calorimetry study to follow Experiment 11

    Design and carry out an experiment to measure the energy content of wood. (A suggested source of wood is the wood splints that are commonly available in chemistry laboratories.) First identify the experimental challenges and devise a way to solve them.

4. A calorimetry study to follow Experiment 11

    Using nuts as an example of a high-calorie food, design and carry out an experiment to measure the energy content in nuts and compare it with other fuels. First identify the experimental challenges and devise a way to solve them.

5. A comparison study of volume vs. weight for titrations, to follow Experiment 13, 14, 15, or 16

    Using a titration analysis of your choice, design and carry out an experiment to compare titrations done by volume (i.e., by counting drops) with titrations done by weight (i.e., by weighing the pipets on a laboratory balance). What are the advantages and disadvantages of each method?

6. A study of melting points to follow Experiment 26

   Design and carry out an experiment to test whether melting point is an accurate measure of purity. Starting with pure samples of aspirin and salicylic acid, determine how melting point varies in mixtures and whether melting point can be used to determine composition of such a mixture.

7. A density of solutions study to follow Experiment 28

   Develop and test a procedure for measuring the alcohol content in an unknown alcohol-water mixture. Pure alcohol has a density that is lower than that of water.

8. A density of solutions study to follow Experiment 28

   Devise and test a procedure for measuring the ethylene glycol content in automobile radiator coolant. Pure ethylene glycol has a higher density than water. (Most automobile coolant systems contain ethylene glycol mixed with water. This prevents freezing in winter as well as allowing the engine to run at a higher temperature than with pure water.)

9. A study of vitamin C stability to follow Experiment 29

   Design and carry out an experiment to test whether vitamin C in orange juice decomposes during storage and whether temperature makes a difference.